Ecological Studies, Vol. 164

Analysis and Synthesis

Edited by

I.T. Baldwin, Jena, Germany
M.M. Caldwell, Logan, USA
G. Heldmaier, Marburg, Germany
O.L. Lange, Würzburg, Germany
H.A. Mooney, Stanford, USA
E.-D. Schulze, Jena, Germany
U. Sommer, Kiel, Germany

Ecological Studies

Volumes published since 1997 are listed at the end of this book.

Springer

Berlin
Heidelberg
New York
Hong Kong
London
Milan
Paris
Tokyo

R. Wirth   H. Herz   R.J. Ryel   W. Beyschlag   B. Hölldobler

Herbivory of Leaf-Cutting Ants

A Case Study on *Atta colombica* in the Tropical Rainforest of Panama

With 82 Figures, 16 in Color, and 32 Tables

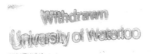

 Springer

Dr. Rainer Wirth
Universität Kaiserslautern, Fachbereich Biologie, Lehrstuhl für Allgemeine Botanik,
Postfach 3049, 67653 Kaiserlautern, Germany

Dr. Hubert Herz
Universität Würzburg, Theodor-Boveri-Institut für Biowissenschaften,
Lehrstuhl für Verhaltensphysiologie und Soziobiologie (Zoologie II),
Am Hubland, 97074 Würzburg, Germany

Prof. Dr. Ronald J. Ryel, Department of Forest, Range and Wildlife Sciences,
Utah State University, 5230 Old Mail Hill, Logan, Utah 84322-5230, USA

Prof. Dr. Wolfram Beyschlag
Universität Bielefeld W4–107, Lehrstuhl für experimentelle Ökologie und Ökosystembiologie,
Universitätsstr. 25, 33615 Bielefeld, Germany

Prof. Dr. Bert Hölldobler
Universität Würzburg, Theodor-Boveri-Institut für Biowissenschaften,
Lehrstuhl für Verhaltensphysiologie und Soziobiologie (Zoologie II),
Am Hubland, 97074 Würzburg, Germany

Cover illustration: by Dr. Damond Kyllo, University of Missouri, USA. The illustration is a visual homage to the leaf-cutting ants and their tropical rainforest environment as perceived by both the artist and authors after many years of living and working on Barro Colorado Island.

ISSN 0070-8356
ISBN 3-540-43896-3 Springer-Verlag Berlin Heidelberg New York

Library of Congress Cataloging-in-Publication Data

Herbivory of leaf-cutting ants: a case study on Atta colombica in the tropical rainforest
of Panama / R. Wirth ... [et al.].
    p. cm. -- (Ecological studies, ; v. 164)
  Includes bibliographical references (p.).
  ISBN 3540438963 (alk. paper)
    1. Atta colombica--Ecology--Panama--Barro Colorado Island. 2. Rain forest
ecology--Panama--Barro Colorado Island. 3. Insect-plant relationships--Panama--Barro
Colorado Island. I. Wirth, R. (Rainer), 1964- II. Series

QL568.F7 H47 2003
595.79'6--dc211   2002030473

Springer-Verlag Berlin Heidelberg New York
a company of BertelsmannSpringer Science+Business Media GmbH

http://www.springer.de

© Springer-Verlag Berlin Heidelberg 2003
The copyright of the original illustrations (drawings or photographs) remains with the authors.

Printed in Germany

The use of general descriptive names, registered names, trademarks, etc. in this publication does not imply, even in the absence of a specific statement, that such names are exempt from the relevant protective laws and regulations and therefore free for general use.

Cover design: *design & production* GmbH, Heidelberg
Typesetting: Kröner, Heidelberg

SPIN 10691950   31/3150 YK – 5   4   3   2   1   0 – Printed on acid free paper

For Edward O. Wilson,
premier myrmecologist,
naturalist, and ardent advocate
for biodiversity research,

in appreciation for his many pioneering
contributions to our understandig
of leaf-cutting ant societies.

Preface

The idea for this study was born at the University of Würzburg, Germany, during the time when two of the authors (W.B. and R.J.R.) were working on the effects of needle loss of declining red spruce trees on the light microenvironment within the forest canopy. In this system it could be shown that the partial loss of foliage led to a significant increase in photosynthetically active radiation within the canopy which then benefited the remaining healthy foliage, particularly in dense canopies. Since Würzburg is one of the places in Germany where interdisciplinary biological research has a long tradition, these results were also discussed with zoologists. One of them (B.H.) became particularly interested because one of his major research interests was leaf-cutting ants, an organism which creates substantial foliage losses in the canopies of Neotropical rainforests. Quickly it became obvious that by combining the experience of both botanists and zoologists it would be possible to quantitatively assess the impact of these selective herbivores on their natural ecosystem, an important question which at that time had never been addressed before, although leaf-cutting ants, because of their importance as pests, certainly belonged to the most studied insect species of the Neotropics. As a starting hypothesis, it was formulated that leaf-cutting ants, by continuously creating small gaps in a rainforest canopy where light is typically a major limiting resource, would allow light to penetrate deeper into the canopy, thereby providing opportunities for species of the middle and lower canopy and, thus, perhaps contribute to the high biodiversity. Very quickly it became clear that testing this hypothesis was not an easy task to achieve, because such ant-induced changes in the canopy light climate could, of course, not be analyzed without substantial quantitative information on rainforest canopy structure, distribution and activities of the ants, herbivory rates, and the response of the affected plants to this herbivory. Furthermore, after a closer look, it became clearer that leaf-cutting ants do affect much more than just the light resource, and leaf-cutting ant-induced disturbance occurs at multiple scales – from the single plant all the way up to the ecosystem. Unfortunately, information on the spatiotemporal herbivory patterns of leaf-cutting ants was scarce and long-term studies on this subject were practically missing. Hence, more and more dimensions and scales had to be included into our considerations and finally we ended up designing a major integrated and interdisciplinary ecosystem research project. Not surprisingly, after the fieldwork had started, unexpected results (for instance, the high frequency of ant colony movements) added even more aspects, and what was initially planned as a 2–3-year study, became a major enterprise lasting a total of 8 years, including 5 years of fieldwork.

Nevertheless, looking at the results, we can say that working on this project for such a long time not only answered quite a few of our questions and, thus, increased our knowledge of the ecological role of leaf-cutting ants, but most importantly also provided fundamental insights into the functioning of the rainforest ecosystem for each of us. We hope by writing this volume we were able to transport not only our scientific results and ideas (many of them originating from the doctoral theses of R.W. and H.H.), but also our excitement about these fascinating animals and their extraordinary habitat.

This study would not have been possible without the help and the input of quite a few people and organizations. First, we would like to thank the Smithsonian Tropical Research Institute for permission to use their research station on Barro Colorado Island in Panama and we gratefully acknowledge the help of their administration in Balboa and the staff of the research station. In particular, we would like to thank the station's scientific coordinator Mrs. Oris Acevedo for her incredible support. We are grateful for all the help and stimulation of the STRI staff scientists, particularly Dr. Egbert Leigh and Dr. Klaus Winter. Further, Mr. Eduardo Sierra †, Mr. Rolando Perez, and their colleagues contributed their valuable experience in determining plant species (particularly from small harvested fragments). The following student workers and graduate students were involved in the fieldwork and contributed valuable data material: Jürgen Berger, Eva Cué-Bär, Querube D. Fuenmayor, Ralf Hömme, Dorkas Kaiser, Boris Leymann, Mir Rodriguez Lombardo, Andreas Meier, Stefanie Rottenberger, Gerold Schmidt, Antje Siegel, and Alexandra Weigelt. Many ideas would never have been created without the inspiring discussions and the valuable input of the following scientists: Dr. Elisabeth Coley, Dr. Cameron Currie, Dr. William Eberhard, Dr. Bettina Engelbrecht, Dr. Allen Herre, Dr. Jerry Howard, Dr. Tom Kursar, Dr. Egbert Leigh, Dr. Flavio Roces and Dr. Thomas Seeley (thanks to both for critically reading the natural history chapter), Dr. Thomas Steinlein, and Dr. Mary Jane West-Eberhard. Dr. Edward O. Wilson gave us permission to use some of the material from publications of which he was author or coauthor. Further, we thank Dr. Damon Kyllo who did a great job in designing the cover picture and Mr. Christian Ziegler for permission to publish some of his excellent photographs. Technical and administrative help in Germany was provided by Doris Faltenbacher-Werner, Elke Furlkröger, Evelyn Hatzung, Ursula Herlth, Monika Noak, Christine Schlüter, and Dr. Hans Zellner. The project was funded by the Deutsche Forschungsgemeinschaft (SFB 251; Ökologie, Physiologie und Biochemie

pflanzlicher und tierischer Leistung unter Streß, University of Würzburg).
Last, but not least, we would like to express our gratitude to Dr. Otto Ludwig
Lange for inviting us to publish this monograph in the Ecological Studies
Series and for his continuous and stimulating support during the preparation
of the manuscript.

September 2002   *The Authors*

Contents

1 About This Book

Plant–animal interactions have become a major focus of ecological research during recent years. One field of major interest in this connection is the process of herbivory. Interactions between plants and herbivorous insects constitute an important component of almost any ecosystem. These interactions occur at low trophic levels and as a result often influence food webs. They play a crucial role in the recycling of matter and, hence, energy and nutrient flows. Last, but not least, these interactions evolve and coevolve, and are considered to be one of the processes and driving forces which organize ecosystems (Crawley 1983; Zwölfer 1987). Generally, insect herbivory has been shown to be one of the disturbance effects which can positively influence secondary plant succession, and thus, species diversity (McBrien et al. 1983; Brown and Gange 1990, 1992; Pacala and Crawley 1992; Davidson 1993; Vasconcelos and Cherret 1997; see Chap. 15).

Due to their immense importance as pests (see below), leaf-cutting ants are probably the most studied tropical insects (Weber 1972a,b; Hölldobler and Wilson 1990; Fowler et al. 1990). Cherrett and Cherrett (1975) compiled 1020 references and this number has continuously increased since then. Naturally, the bulk of work on leaf-cutting ants has been devoted to research on pest control (for review see Vander Meer et al. 1990) and to related topics such as the selection of host plants (Cherrett 1968; Rockwood 1975; for review see Vasconcelos and Fowler 1990), including the elucidation of the chemical and physical aspects of the preferred host plants (e.g., Hubbell et al. 1984; Howard 1988; Howard et al. 1989; Nichols-Orions 1991a–c). Further, the phenomenon that leaf-cutting ants distribute their foraging activities over a large area, instead of using the vegetation immediately around their nest site, has provoked a long-lasting discussion on foraging strategies (Cherrett 1968; Fowler and Styles 1980; Rockwood and Hubbell 1987; Shepherd 1982). From the ecological viewpoint, the most striking aspect of leaf-cutting ants in the rainforest ecosystem is their importance as generalist herbivores (see Chap. 15). Of all herbivore groups of comparable taxonomic diversity, the leaf-cutting ants of the genera *Atta* and *Acromyrmex* are the most dominant herbivores in the tropics of the New World (Wilson 1986). Wint (1983), for example, estimated that *Atta* species were responsible for some 80 % of visible leaf damage found in Panamanian rainforests (but see also Cherrett 1989 and Chap. 11, this Vol.). The ecological and economic impact of the symbiosis between leaf-cutting ants and their plant-consuming mutualist fungus is strikingly revealed by the fact that the two genera *Atta* and *Acromyrmex* are the most serious agricultural pests of tropical and subtropical America, causing enormous economic damage (several thousand million dollars annually) to the neotropic agricultural industry (cf. Cherrett 1986).

Despite considerable research efforts, there are still major aspects of the biology and ecology of leaf-cutting ants that have received little attention and hence have been rarely addressed in the extensive review literature. For instance, there is minimal information about the origin, maintenance and persistence of foraging trails and their spatial and temporal variability (Shepherd 1985; Farji Brener and Sierra 1993; Howard 2001). Also, virtually nothing is known about the survival rates of colonies or their maximum ages under field conditions (Autuori 1941; Fowler et al. 1986b). While the rate of colony turnover is critical for assessing the importance of leaf-cutting ants in rainforest ecosystems, our knowledge of the temporal variation of populations of colonies is correspondingly limited (Perfecto and Vandermeer 1993). Finally, and most surprisingly, detailed studies on the impact of these polyphagous herbivores on rainforest plants and community dynamics are very scarce (but see Vasconcelos 1997, 1999; Vasconcelos and Cherrett 1996; Rao et al. 2001). Although various attempts have been made to estimate the amounts of vegetation carried into the nests of leaf-cutting ants (Hodgson 1955; Cherrett 1968; Rockwood 1975; Blanton and Ewel 1985; Haines 1978; Lugo et al. 1973), no simultaneous estimates have been made of the production of the forest area impacted or the affected standing leaf crop. Little has been done to ascertain the consequences of the foliage loss in these canopies. Most importantly, long-term studies of the quantitative and qualitative effects of the leaf-cutting ant harvesting activities are completely lacking. The case study presented in this volume was undertaken to address most of these open questions.

Starting with a general description of the natural history of leaf-cutting ants in Chapter 2, we subsequently present the results of a 5-year-long case study carried out from 1993 to 1998 in a semideciduous tropical rainforest in Panama. Chapters 3–6 focus on the habitat characteristics of the ants by first describing the study area (Chap. 3) and the floristic composition of the forest (Chap. 4), followed by detailed analyses of the forest light regime (Chap. 5) and the canopy structure (Chap. 6). In the following chapters, the ants and their activities are the center of interest. Information is given on the spatial and temporal colony distribution (Chap. 7), the short-term and long-term harvest dynamics (Chap. 8), the trail system and the foraging area of the colonies (Chap. 9) and the characteristics of their host plant selection (Chap. 10). Subsequently, we analyze the impact of the ants on the rainforest ecosystem starting with the calculation of herbivory rates at several scales

(Chap. 11) and following up with an analysis of the effects on the light climate in various parts of the canopy (Chap. 12). The next two chapters discuss the importance of the ants as seed dispersers (Chap. 13), and assess the role of the ants in nutrient cycling and the water relations of harvested plants (Chap. 14). Integrating the information gained from this study, we finally discuss the role of leaf-cutting ants as a mechanism which may help to maintain ecosystem biodiversity (Chap. 15).

This compilation of leaf-cutting ant activity and interactions with the rainforest vegetation on Barro Colorado Island (see Plate 1) represents an appropriate model for summarizing and extending knowledge about herbivorous insect-plant relationships, and assessing the resulting consequences on structural and functional attributes of tropical ecosystems. To the knowledge of the authors, a study of similar breadth and duration on this subject has not been published. Therefore, we hope that this synthesis volume on the effects of leaf-cutting ant herbivory at multiple scales will act as a reference for researchers and land managers working in the fields of plant-animal interaction, herbivory, community ecology and biodiversity.

2 The Natural History of Leaf-Cutting Ants

2.1 The Significance of Social Insects

No one can give us an exact number of animal species living on earth today, but all biologists agree that millions more species exist than the approximately 1.5 million that have been described so far. Quantitative faunistic investigations in many habitats suggest about 8 million extant species; other assessments claim 30 million species or even more (Wilson 1992 from Erwin 1982, but see Novotny et al. 2002). Most of these species that share mother earth with us are still unknown to science, and sadly, may never become known, because of ongoing man-made habitat destruction and ensuing species extinction.

About half of all described animal species belong to the class Insecta (~750,000 described insect species), of which only about 2 % live in eusocial systems. We consider an insect society to be eusocial when the following criteria are met: cooperative care for the immature individuals, overlap of at least two generations in the same society, and existence of reproductive and non-reproductive individuals.

In an evolutionary advanced grade of eusociality, we find distinct morphological castes: a reproductive caste (queen), and a nonreproductive caste (worker). The latter can be further subdivided into several morphologically distinct subcastes, such as minor, media, and major workers. In advanced eusocial societies, workers exhibit a sophisticated system of division of labor, whereby task- and role-assignments can depend on the age of the individual (age polyethism) or on morphological features (physical polyethism), or both. Depending on the species-specific social organization and ontogenetic status of the colony (society), and (or) on the particular situation of the society, the distribution of roles and tasks among the workers can be more or less flexible. Nevertheless, a good "rule of thumb" is that young individuals spend most of their time inside the nest, being mainly engaged in the care of the queen and brood, whereas older workers are more involved with activities outside the nest, such as midden work, nest construction, foraging, and defense of nest and territory. This remarkable cooperation and division of labor bestows a tremendous advantage upon the social insects. Whereas at any given moment a solitary organism can be doing only a few things and can be in only one place, an insect society can perform many activities and can be in several different places by deploying its worker force. This is most likely the reason why social insects play such a dominant role in most terrestrial ecosystems.

Although only 2% of the known insect species are eusocial, social insects dominate the terrestrial environments around the world. For example, in a much cited study, Fittkau and Klinge (1973) found in the Brazilian Terra Firme forest that ants and termites (all species of which live in societies) together compose roughly 30% of the animal biomass. If we add the stingless bees and polybiine wasps, then altogether the social insects make up more than 75% of the entire insect biomass. Based on these and other data, we agree with the following statement by E.O. Wilson (1990): "It is my impression that in another, still unquantified sense these organisms, and particularly the ants and termites, also occupy center stage in the terrestrial environment. They have pushed out solitary insects from the generally most favorable nest sites. The solitary forms occupy the more distant twigs, the very moist or dry excessively crumbling pieces of wood, the surface of leaves – in short, the more remote and transient nesting places. .... At the risk of oversimplification, the picture we see is the following: social insects are at the ecological center, solitary insects at the periphery."

The ants (Formicidae, Hymenoptera) are divided into 16 subfamilies (Bolton 1994). They exhibit the most impressive adaptive radiation of any of the eusocial insect groups. Close to 10,000 species are known to science, but based on the rate of discovery of species, systematists estimate that approximately 20,000 species exist. Although all ant species are eusocial, the social organizations of particular species groups vary greatly. For example, in some species, each colony has only one queen (monogynous), while in others each colony has many queens (polgynous). Some species form colonies that consist of a relatively small number of workers (50–200), while other species form colonies containing hundreds of thousands or even millions of workers. Equally diverse are the modes of colony foundation and colony reproduction, the ways in which the division of labor systems are organized, how communication among the individuals and groups of individuals function, and how the colonies forage and which resources they exploit (see Hölldobler and Wilson 1990).

Myrmecologists are intrigued by several pinnacles in ant evolution: the army ants of the Neotropics and the driver ants of Africa; the weaver ants of the genus *Oecophylla* of Africa, Asia and Australia; the "supercolonies" of the ant *Formica yessensis* on the Ishikari Coast of Hokkaido, each of which is composed of 306 million workers and 1,080,000 queens living in 45,000 interconnected nests across a territory of 2.7 km² (Higashi and Yamauchi 1979); the supercolonies of *Formica lugubris* of Alpine regions (Cherix and Bourne 1980); the migrating herdsmen of the genus *Hypoclinea* in the rainforest of the Malaysian peninsula (Maschwitz and Hänel 1985); and last, but not least, the fungus growers of the myrmicine tribe Attini. This is by no means a complete list of the astounding diversity of life styles in the Formicidae; it is merely a sample of the most spectacular ways of life the ants have "invented" in the course of their approximately 100 million years evolutionary history.

2.2 The Fungus Growers

Members of the myrmicine tribe Attini share with macrotermitine termites and certain wood-boring beetles the sophisticated habit of culturing and eating fungi. The Attini are a morphologically distinctive group limited to the New World, and most of the 12 genera and more than 200 species occur in the tropical regions of Mexico and Central and South America, though a few species live in the southern portions of the United States of America, and some have even adapted to more arid habitats in the southwestern states. One species, *Trachymyrmex septentrionalis*, even ranges north to the pine barrens of New Jersey, while in the opposite direction several species of *Acromyrmex* penetrate to the cold temperate deserts of central Argentina (Hölldobler and Wilson 1990). Phylogenetic analysis of the fungus growing ants by Schultz and Meier (1995) and Wetterer et al. (1998) confirm the monophyly of the Attini, and support the hypothesis that fungus growing in attine ants has evolved only once. The studies further provide evidence that the neotropical genus *Blepharidatta* is the sister group of the Attini (Diniz et al. 1998).

Fungus farming by ants of the tribe Attini originated in the early tertiary, whereas human agriculture arose only around 10,000 years ago. Thus the ants' agricultural life style predates that of humans by some 50 million years (Mueller et al. 1998). Most cultivated fungi belong to the basidiomycete family Lepiotaceae (Agaricales: Basidiomycola), and the great majority of attine fungi belong to two genera, *Leucoagaricus* and *Leucocoprinus* (Leucocoprineae) (Chapela et al. 1994; Mueller et al. 1998; Johnson 1999). Mueller et al. (2001) reason that because "most basal attine lineages cultivate leucocoprineous mutualists, attine fungiculture likely originated with the cultivation of leucocoprineous fungi". A remarkable exception exists: some species of the genus *Apterostigma* have secondarily changed to nonlepiotaceous fungi that belong to the family Tricholomataceae (Chapela et al. 1994; Mueller et al. 1998). However, with the exception of these tricholomataceous mutualists, attine ants associate entirely with a closely related group of leucocoprineous fungi (Mueller et al. 2001).

It has long been assumed that fungal transmission is strictly vertical, i.e., a transfer of fungal cultivars from parent to offspring nests (Weber 1972a). This would imply that the clonally propagated fungal lineages evolved in parallel with the lineages of the ant mutualists over millions of years. However, at least some "lower" attines (phylogenetic basal lineages) propagate cultivars that were recently domesticated from free-living populations of Lepiotaceae (Mueller et al. 1998), but the "higher" attines (derived lineages) are thought to propagate ancient clones several million years old (Chapela et al. 1994). How ancient these clones really are remains uncertain. Mueller et al. (1998) contend that propagation from parent to daughter colonies "may indeed be the general rule for most attine ant species over short-term evolutionary time spans, but the long-term evolutionary histories of attine fungal lineages may be complex, involving both lateral transfer between distantly related ant lineages and repeated cycles of domestication of free-living fungi followed by a return to the free-living state." (also see Bot et al. 2001b). For the various hypotheses explaining the origin of the attine ant-fungus mutualism, we refer to the excellent review by Mueller et al. (2001), who, after carefully evaluating the seven prevalent hypotheses that have been proposed by different authors, conclude: "... the attine ant-fungus mutualism probably arose from adventitious interactions with fungi that grew on walls of nests built in leaf litter (Emery 1899), or from a system of fungal myrmechory in which specialized fungi relied on ants for dispersal (Bailey 1920), and in which the ants fortuitously vectored these fungi from parent to offspring nests prior to a true fungicultural stage."

The Attini are an enormously successful group. In the vast subtropical and tropical zones of the New World, attines are among the dominant ants. Many of the species gather pieces of fresh leaves and flowers to nourish the fungus gardens, and the species in the genera *Atta* and *Acromyrmex* rely on this source exclusively (see Plates 2–4, 6 above, 7 above). They are the leaf-cutting ants in the true sense. Since they attack most kinds of vegetation, including crop plants, they are serious economic pests. The leaf-cutting ants of the genera *Atta* and *Acromyrmex* were preadapted for their role as agricultural pests by their ability to use many plant species with the aid of their symbiotic fungi, which serve as a sort of ancillary digestive system (Plate 6, 7, 16). The ants can build up high population densities, such as 5 colonies/ha in *Atta vollenweideri*, with each colony containing a million or more workers (Fowler et al. 1986a,b). Because many species thrive in cleared land and secondary forests, leaf-cutting ants have benefited by the spread of European civilization to the New World tropics. The ubiquitous *Atta cephalotes*, for example, is specialized to live in forest gaps, thus it is able to invade subsistence farms and plantations from Mexico to Brazil (Cherrett and Peregrine 1976).

While it is true that leaf-cutting ants cause serious problems, it would be a mistake to think of them solely as pest species. Leaf-cutting ants are an integral part of the ecosystems of the New World tropics and warm temperate zones. In fact, they supplant to a large extent the populations of herbivorous mammals, which are relatively sparse through most of the New World tropics. They prune the vegetation, stimulate new plant growth, break down plant material, and enrich the soil. As shown later, they may also play a role in maintaining the biodiversity of their ecosystem (see Chap. 15)

2.3　The Natural History of Leaf-Cutting Ants of the Genus *Atta*

In phylogenetic studies of fungus-growing ants, the two genera of the so-called leaf-cutting ants, *Acromyrmex* and *Atta*, are combined with three additional genera in the derived, monophyletic group of the "higher attines" (Weber 1972b). The remaining seven genera are assembled in the "lower attines" (Schultz and Meier 1995; Wetterer et al. 1998). Species of the lower attines do not cut and use leaves as the main cultivating substrate for their symbiotic fungus, but rather collect a large variety of dead vegetable matter, insect frass or other kinds of organic material (including plant seeds, fruits and insect corpses; Leal and Oliveira 2000). In fact, based on quantitative studies of the foraging habits and other ecological data, Leal and Oliveira suggested a somewhat different grouping of the Attini that nevertheless agrees with the phylogeny proposed by Schultz and Meier. They divide the Attini into three groups. The most "primitive" group comprises the genera *Cyphomyrmex, Mycetarotes, Mycocepurus, Myrmicocrypta, Apterostigma* and *Mycetosoritis*; the intermediate group contains the genera *Sericomyrmex, Trachymyrmex, Mycetophylax*; and the advanced group would be formed by the leaf-cutting ants *Acromyrmex* and *Atta*, together with the genus of parasitic ants *Pseudoatta*.

Of the two true leaf-cutting ant genera, *Acromyrmex* has the greater species diversity (24 species, 35 subspecies), whereas *Atta* comprises only 15 species (Bolton 1995). *Atta* species are among the most advanced of all the social insects. The life history traits of the *Atta* species are basically very similar, which permits us to present a general description of their natural history.

2.3.1　Colony Foundation in *Atta*

Each mature *Atta* colony consists of one queen and hundreds of thousands or even millions of workers. Every year, each mature colony produces young reproductive females and males, the alates, which depart from the parental colony on mating flights or so-called nuptial flights. The nuptial flights of all the *Atta* colonies belonging to the same species and living in the same habitat appear to be well synchronized. In *Atta sexdens* of South America, for example, they take place in the afternoon anytime from the end of October to the middle of December, while in *A. texana* of the southern United States they occur at night (Autuori 1956; Moser 1967). Mating apparently takes place high in the air, and since many colonies conduct their nuptial flights during the same period of the day, the probability of outbreeding is high. Although mating has never been observed in nature, based on sperm counts taken from the spermathecae of newly mated *A. sexdens* queens, Kerr (1962) proposed that individual queens are inseminated by at least three to eight males. These findings were later confirmed in studies employing DNA analyses (Corso and Serzedello 1981; Fjerdingstad and Boomsma 1997, 1998; Fjerdingstad et al. 1998; see also Boomsma and Ratnieks 1996). For example, Fjerdingstad and Boomsma (1997) found that in *A. colombica* the average number of fathers per colony is almost three, however, variation in paternity shared among fathers means that the effective paternity frequency is only two. The biological significance of multiple paternity in *Atta* is not yet entirely clear. Mainly three hypotheses are being discussed in the literature: (1) the polyandry-for-sperm hypothesis (Cole 1983). Colonies of *Atta* species are extremely populous and they have a long life span (10–15 years or even longer), with a single queen as sole reproductive. Since the queen mates only once in her entire life, she has to stock up a huge sperm supply. In Kerr's study, the estimated numbers of sperm varied among the queens from 206 to 320 million, seemingly more than enough to last an individual 10 or more years. (2) The genetic-diversity-disease-resistance hypothesis (Hamilton 1987; Sherman et al. 1988). High within-colony genetic diversity may confer a fitness advantage, for instance in relation to disease resistance. (3) High within-colony genetic diversity may have a positive effect on worker task efficiency (Crozier and Page 1985). No empirical evidence exists yet that favors one of the three hypotheses (Fjerdingstad and Boomsma 1998).

After the mating flight all males die. The sole function of ant males is to provide sperm, which are stored in the spermathecae of the queens. Thus the life spans of male ants (which develop from uninseminated eggs and are therefore haploid) are very short. However, because of the long preservation time of sperm in the queens' "sperm banks", ant males can become fathers many years after they have died. Mortality is also very high for the queens, especially during the mating flight and immediately afterward, as the queens attempt to start new colonies. According to Dix and Dix (unpubl., cited in Fowler et al. 1986b), mortality for *A. cephalotes* queens during mating flights may be as high as 52%. Out of 13,300 newly founded colonies of *A. capiguara* in Brazil, only 12 were alive 3 months later (Fowler et al. 1986b). From a start of 3558 incipient *A. sexdens* colonies, only 90 (2.5%) were alive after 3 months (Autuori 1950). In another study only 10% of *A. cephalotes* colonies survived the first few months after colony foundation (Fowler et al. 1986b).

Before departing on her mating flight, each *Atta* queen packs a small wad of mycelia of the symbiotic fungus into her infrabuccal pocket (cibarium), a cavity located beneath the opening of the esophagus. Following the nuptial flight, the queen casts off her wings and excavates a nest chamber in the soil. This incipient nest consists of a narrow entrance gallery which descends 20–30 cm to a single chamber (approx. 6 cm long; Fig. 1). The queen now spits out the mycelial wad which serves as an inoculum to start a new fungus garden. By the third day fresh mycelia have begun to grow, and the queen has laid three to six eggs (Autuori 1956; Hölldobler and Wilson 1990). At the end of the

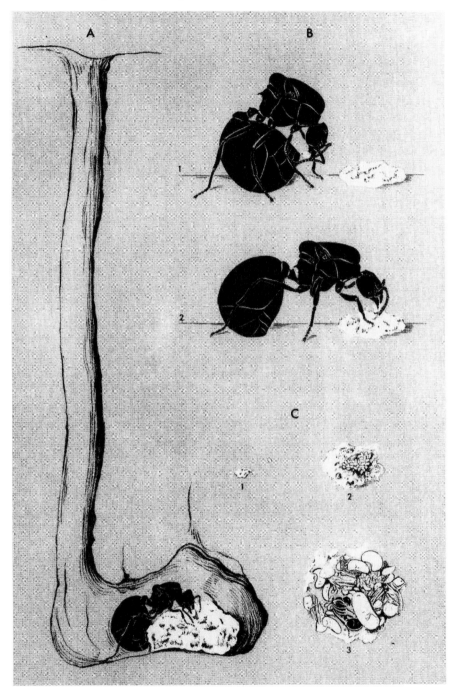

Fig. 1. Colony founding in *Atta*. **A** A queen in her first chamber with the beginning fungus garden; **B** the queen fertilizes the garden by freeing a hyphal clump and applying an anal droplet to it; **C** three stages (*1–3*) in the concurrent development of the fungus garden and first brood. (Wilson 1971; drawing by T. Hölldobler-Forsyth)

first month the brood, now consisting of eggs, larvae, and perhaps pupae, is embedded in the center of a mat of proliferating fungus. During this initial phase of colony foundation the queen cultivates the fungus garden herself, mainly by fertilizing the garden with fecal liquid. The queen consumes 90 % of the eggs she lays. When the first larvae hatch, they are also fed with eggs. Apparently, the queen does not eat from the initial fungus culture which initially is very fragile. If the queen fails to build up a healthy fungus garden, the whole colony founding process is doomed. Instead, the queen subsists entirely on her own fat body reserves and on catabolizing the wing muscles, which are no longer needed.

When the first workers enclose, they begin to feed on the fungus and they take over the fungus culture activities. The egg laying rate of the queen increases: Not all her eggs are viable, for some are large trophic eggs formed in the oviduct by the fusion of two or more distinct but ill-formed eggs (Bazire-Benazet 1957). These are given by workers to developing larvae. After a week or so the young workers open the clogged nest entrance and start foraging on the ground in the immediate vicinity of the nest. They collect bits of leaves from which they prepare inside the nest a humus on which the fungus is cultivated. The queen ceases attending to the brood and fungus garden; she becomes an egg laying machine for the rest of her long life. The workers take over all "somatic" duties: foraging, caring for fungus garden, raising the brood, extending the nest structures, and defending the colony against predators and competitors (Plate 6).

The fungus cultivated by *Acromyrmex* and *Atta* species produces hyphal-tip swellings, called "gongylidia", and the densely packed clusters of these are called "staphylae" (Plate 7 below). They are easily plucked by the ants and eaten or fed to the larvae. The structures are rich in lipids and carbohydrates, while the hyphae are richer in proteins (Quinland and Cherrett 1979; Febvay and Kermarrec 1983; for review see Mueller et al. 2001). When given a choice in feeding experiments *Atta* workers prefer staphylae over hyphae (Quinland and Cherrett 1978, 1979; Angeli-Papa and Eymé 1985).In addition, *Atta* workers live longer when feeding on staphylae than on hyphae (Bass and Cherrett 1995). Thus the gongylidia bodies appear to contain the best balanced blend of nutritional components, which is not yet entirely understood (Mueller et al. 2001).

As fresh leaves and other plant cuttings are brought into the nest (Plates 2 , 4, 7 above), they are subjected to a process of degradation before being inserted into the garden substratum (for details see Hölldobler and Wilson 1990). The ants subsequently pluck tufts of mycelia from other parts of the garden and plant them on newly formed portions of the substratum. The transplanted mycelia grow as much as 13 µm/h.

However, the fungus appears not to be the only source of nutrition for leaf-cutting workers. Littledyke and Cherrett (1978) reported that *Atta* and *Acromyrmex* workers feed directly on plant sap. The sap must be crucial to the workers, because Quinland and Cherrett (1979) found that only 5 % of their energy requirements are met by ingestion of the contents of fungal staphylae. In contrast, the larvae are able to subsist and grow entirely on the staphylae. The queen appears to obtain at least a substantial part of her food from trophic eggs laid by workers and fed to her at frequent intervals.

The growth of the incipient colony is slow in the first 2 years. During the next 3 years, it accelerates quickly and tapers off as the colony starts to produce winged males and queens. The ultimate size reached by colonies of *Atta* is enormous. For example, one *A. sexdens* nest, more than 6 years old, contained 1920 chambers of which 248 were occupied by fungus garden and ants. The loose soil that had been brought out and piled on the ground by the ants during the excavation of their nest weighed approximately 40,000 kg (40 tons)! As these numbers suggest, the population size of mature colonies of *Atta* must be tremendous. Fowler et al. (1986b) summarized the information from several publications: the numbers of workers in a single colony has been estimated as 1–2.5 million in *Atta colombica*, 3.5 million in *A. laevigata*, 5–8 million in *A. sexdens*, and 4–7 million in *A. vollenweideri* (Fig. 2)

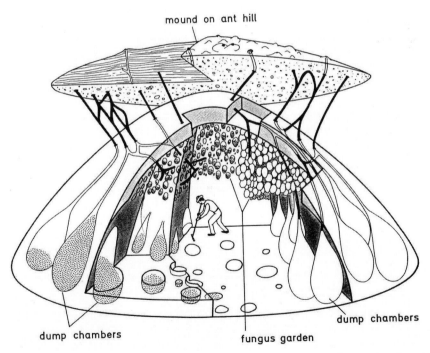

mound on ant hill

dump chambers

dump chambers

fungus garden

Fig. 2. The plan of a mature nest of the leaf-cutting ant *Atta vollenweideri*, based on actual excavations. The upper mound of soil was brought to the surface by the ants during the digging of the nest. The dump chambers contain exhausted substrate. The fungus is cultured in the fungus garden chambers. (Hölldobler and Wilson 1990, modified from J.C.M. Jonkman, in Weber 1979)

2.3.2 Division of Labor in *Atta* Colonies

The most thorough studies on the division of labor and worker caste-systems in *Atta* were conducted by E.O. Wilson. We will therefore base our brief account mainly on his work with *A. sexdens* (Wilson 1980a,b) and *A. cephalotes* (Wilson 1983a,b, 1985). Presumably, the pattern is similar in the other *Atta* species.

Atta leaf-cutting ants have a broad array of physical subcastes in the worker groups. In *Atta sexdens*, for example, the head width varies 8-fold and the dry weight 200-fold from the smallest minor-workers to the huge major-workers (Fig. 3). However, developing colonies, started by a single queen, have a nearly uniform size frequency distributed across a relatively narrow head width

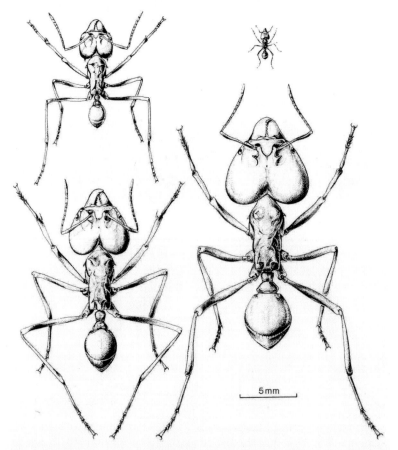

Fig. 3. The caste systems of leaf-cutting ants of the genus *Atta* are among the most complex in the social insects, involving extreme size variation accompanied by strong diphasic allometry. The workers illustrated here are from a single colony of *A. laevigata*. (Hölldobler and Wilson 1990; drawing by T. Hölldobler-Forsyth)

range of 0.8–1.6 mm. The reason for this is that workers in the span 0.8–1.0 mm are required as gardeners of the symbiotic fungus, whereas workers with a head width of 1.6 mm are the smallest that can cut vegetation of average toughness. This range also embraces the worker size groups most involved in brood care. Thus, the queen produces about the maximum number of individuals who together can perform all the essential colony tasks. As the colony continues growing, the worker size variation broadens in both directions, to head width 0.7 mm or slightly less at the lower end to more than 5.0 mm at the upper end, and the frequency distribution becomes more sharply peaked and strongly skewed to the larger size classes. This complex caste system reflects the division of labor in *Atta*, which is closely adapted to the collecting and processing of fresh vegetation for fungal substrate.

The *Atta* workers organize the gardening operation in the form of an assembly line. The most frequent size group among foragers consists of workers with head width 2.0–2.2 mm. At the opposite end of the line, the care of the delicate fungal hyphae requires very small workers, and this task is filled within the nests by workers with head widths predominantly 0.8 mm. The intervening steps in gardening are conducted by workers of graded intermediate size (Fig. 4).

After the foraging medias drop the pieces of vegetation onto the floor of a nest chamber, they are picked up by workers of slightly smaller size, who clip

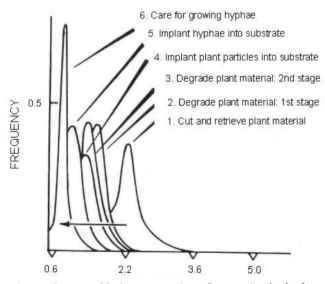

Fig. 4. The assembly-line processing of vegetation by leaf-cutting ants (*Atta sexdens*) depends on an intricate division of labor among the minor and media workers. Medias gather vegetation, after which a succession of ever smaller media and minor workers process it and use it to cultivate fungal hyphae. The polyethism curves are smoothed versions of histograms of head width distributed over 0.1-mm intervals. (Modified from Wilson 1980a)

them into fragments about 1–2 mm across. Within minutes, still smaller ants take over, crush and mold the fragments into moist pellets, and carefully insert them into a mass of similar material. Finally, workers even smaller than those just described pluck loose strands of the fungus from places of dense growth and plant them on the newly constructed surfaces. The very smallest and most abundant workers patrol the beds of fungal strands, delicately probing them with their antennae, licking their surfaces and plucking out spores and hyphae of alien species of mold.

The defense of the colony is also organized to some extent according to size. Though most of the size groups attack intruders, there is in addition a true soldier caste. The extremely large majors have sharp mandibles powered by massive adductor muscles. These giants are especially adept at repelling large enemies. The differential involvement of worker castes during colony defense has been described in a study of *Atta laevigata* (Whitehouse and Jaffé 1996). When the colony is threatened by a potential vertebrate predator, mainly workers belonging to the gigantic soldier caste are recruited. However, when a colony has to defend its nest or foraging area against conspecific or interspecific ant competitors, mainly smaller worker castes respond, which are more numerous and more suitable in territorial combat with enemy ants.

Not all physical worker-subcastes are locked into their task scheme. Wilson (1980a,b) demonstrated for *A. sexdens* that at least three of the four physical castes recognized pass through changes of behavior with aging. Although caste and division of labor in *Atta* are very complex in comparison with other ant systems, they are derived from surprisingly elementary processes of increased size variation, allometry, and alloethism (Wilson 1980a,b). In fact, ant species in general and *Atta* species in particular have been remarkably restrained in the elaboration of their castes. They have relied on a single rule of deformation to create physical castes, which translates into a single allometric curve for any pair of specific dimensions such as head width versus pronotal width (Wilson 1980a,b). Thus, even though *Atta* species possess one of the most complicated caste systems found in ants, it has not evolved anywhere close to the conceivable limit. There are far more tasks than castes: by the first crude estimate seven castes cover a total of 20–30 tasks. Furthermore, one can discern another important phenomenon in *Atta* species that constrains the elaboration of castes: polyethism has evolved further than polymorphism. Ensembles specialized in particular tasks are more differentiated by behavior than by age or anatomy. In the course of evolution *Atta* created its division of labor primarily by greatly expanding the size variation of the workers while adding a moderate amount of allometry and a relatively much greater amount of alloethism (Hölldobler and Wilson 1990).

Alloethism is the regular change in a particular category of behavior as a function of worker size (Hölldobler and Wilson 1990). It stands in close relationship to the phenomenon of task partitioning (Anderson and Ratnieks 2000). "A task is said to be partitioned when it is split into two or more sequential stages and material is passed from one worker to another." This phenomenon is known to myrmecologists from a number of different species in various contexts. Recently, Anderson and Jadin (2001), and Hart and Ratnieks (2001, 2002), paid special attention to its significance in leaf-cutting ants. Here, it can be seen in different circumstances, but appears to be particularly important in leaf-cutting and retrieval of the harvest material, and in the process of waste removal. A typical example is the "bucket brigade" foraging, in which large leafs are cut by some workers and dropped to the ground for further fragmentation and transport to the nest by separate worker crews (Hubbell et al. 1980).

2.3.3 The Maintenance of Fungal Hygiene

The fungus gardens of leaf-cutting ant colonies flourish in subterranean growth chambers with high humidity and tropical temperatures (Plates 6 above, 7 above). It is impressive how apparently nearly pure the ants keep their fungal gardens. They employ a variety of hygienic techniques: the plucking out of alien fungi (Bass and Cherrett 1994; Currie and Stuart 2001) the frequent inoculation of the fungal mycelia onto fresh substrate, the fertilizing of the substrate with enzymes and nutrients (Boyd and Martin 1975), the production of antibiotics to depress competing fungi and microorganisms, and the production of growth hormones. Maschwitz et al. (1970) and Schildknecht and Koob (1970; see also Maschwitz 1974) identified phenylacetic acid, hydroxydecanoic acid (myrmicacin), and indole acetic acid in the secretions of the metapleural glands of *Atta sexdens* workers. They suggest that these compounds play different roles in the purification of the symbiotic fungus culture: phenylacetic acid suppresses bacterial growth, hydroxydecanoic acid inhibits the germination of spores of alien fungi, and indole acetic acid, a plant hormone, stimulates mycelial growth. Recently, a more comprehensive analysis of metapleural gland secretions of *Acromyrmex octospinosus* revealed 20 new compounds, in addition to the already known substances (from *A. sexdens*). They span the whole range of carboxylic acids from acetic acid to long-chain fatty acids, in addition they include some alcohols, lactones and keto acids (Ortius-Lechner et al. 2000).

The prevailing assumption that fungus-culturing ants maintain their fungus gardens in germ-free condition (Bass and Cherrett 1994; North et al. 1997) has to be revised based on growing evidence showing that fungus gardens are often contaminated by bacteria, yeast, and other fungi (Craven et al. 1970; Fisher et al. 1996; Currie et al. 1999a,b; Currie 2001a,b). The ants cannot prevent contamination, but they are able to inhibit the growth of the invading microorganism and foreign fungi. It has been suggested that one of the main countermeasures of the ants against parasitic fungi is to maintain the fungus cultures at a pH of 5, because this is optimal for the growth of the symbiotic fungus (Powell and Stradling 1986), but detrimental for pathogenic fungi. Indeed, it has been demonstrated that the pH rises to 7 or 8 when the ants are removed and within a few days parasitic fungi and bacteria spread rapidly in the fungus cultures (Kreisel 1972). Ortius-Lechner et al. (2000) suggest that one of the main functions of the metapleural gland secretions of *Acromyrmex* and *Atta* workers is to reduce the pH of the leaf material brought into the colony from ca. 7–8 to 5. In addition, it has been shown that each of the acids present in the metapleural glands of leaf-cutting ants can have antibiotic properties. For a detailed discussion of these issues we refer the reader to the publication by Ortius-Lechner et al. (2000).

Recently, some striking new findings concerning the "agricultural pathology" of ant fungus gardens have been reported. Currie et al. (1999a) conducted extensive isolations of nonmutualistic fungi from the gardens of attine ants. They found specialized garden parasites belonging to the microfungus genus *Escovopsis* (Ascomycota: anamorphic Hypocreales; Plate 8 above). These parasites are horizontally transmitted between colonies. *Escovopsis* is highly virulent, able to devastate ant gardens which can in turn destroy the entire ant colony. The genus *Escovopsis* appears to be specialized on fungal gardens of attine ants as it has not been isolated from any other habitat, and it is especially prevalent in *Atta* and *Acromyrmex* colonies. Currie et al. (1999a) explain this phenomenon with the following arguments: "The increased prevalence of *Escovopsis* within the more derived attine genera suggests that the long clonal history of these fungal cultivars, perhaps as long as 23 million years (Chapela et al. 1994), makes them more susceptible to losing the 'arms race' with parasites. By contrast, lower attines routinely acquire new fungal cultivars from free-living sexual populations, leading to greater genetic diversity in the fungal mutualist population. This may account for the apparent lower susceptibility to parasitism of the less derived attine lineages."

How do the fungus-growing *Atta* and *Acromyrmex* cope with this vital threat? Obviously, the successful maintenance of a healthy fungus garden involves a continuous struggle to control the fungal parasite *Escovopsis*. Some deterring effect might come from the metapleural gland secretions, but the main weapon against *Escovopsis* appears to be a third mutualist associated with attine ants, an actinomycete of the genus *Streptomyces* that produces antibiotics that specifically suppress the growth of *Escovopsis* (Currie et al. 1999a,b; Currie 2001a,b). *Streptomyces* inhabit regions of the ants' cuticle that are genus-specific. In *Acromyrmex* and *Trachymyrmex*, for example, it is carried on the laterocervical plates of the propleura. This protecting mutualist is transmitted vertically (from parent to offspring colonies) just as the mutualist fungus is. Because of these and other features, Currie et al. (1999b) argue that the relationship between attine ants and *Streptomyces* is an ancient one, because it is not only especially adapted to fight the parasitic fungus, it also promotes the growth of the fungal mutualist in vitro. In extreme cases of infestation the *Atta* colony may be forced to escape from *Escovopsis* by nest emigration (Herz 2001; Plates 8 below, 9).

We would like to close this section with a quote from Currie et al. (1999a): "Although the ant-fungus mutualism is often regarded as one of the most fascinating examples of highly evolved symbiosis, it is now clear that its complexity has been greatly underestimated. The attine symbiosis appears to be a co-evolutionary 'arms race' between the garden parasite, *Escovopsis*, on the one hand, and the tripartite association amongst the actinomycete, the ant hosts, and the fungal mutualist on the other." For a recent review, see Currie (2001b)

When considering the hygiene in fungus-growing ant colonies we must also discuss at least briefly the problem of "waste management". The refuse produced by the fungus is tremendous. Most *Atta* species have special refuse chambers in their nests in which they pile up the fungal waste; however, *A. colombica* makes an exception, they dispose of their refuse outside the nest (Weber 1972a,b). The refuse is loaded with secondary plant compounds and possibly parasitic fungal mycelia. The ants exhibit a strong avoidance of the waste material. Among local people it has long been known that refuse from *Atta* nests can be used as a powerful repellent against ants including *Atta* species. In fact, Zeh et al. (1999) recently conducted a preliminary study demonstrating that *Atta* refuse put around young plants will protect them from *Atta* herbivory. This, of course, raises the question how the *Atta* workers manage the repulsive fungal waste inside and around their nest.

In a first such study, Anderson and Ratnieks (2000) showed that in *Atta colombica* colonies waste management is organized by task partitioning. Most of the time waste material is taken from the nest and deposited on a cache along the trail to the dump (Plates 10, 11). Other workers collected material from the cache and carried it to the main pile. The authors speculate that the adaptive value of partitioning the task of waste removal is possibly related to "reducing spread of disease and parasites into the colony by separating intra- and extra-nidal workers." This is supported by the findings reported by Bot et al. (2001a) and Hart and Ratnieks (2002). Ants exposed to waste material die at a higher rate and waste is often infected by the fungal parasite *Escovopsis*. The authors also report the interesting observations that the "dangerous tasks" of waste management is mainly performed by older workers, who would soon die anyway. This phenomenon of older workers being more prone to risk their lives is obviously an adaptive trait with respect to colony-level efficiency. This relationship of older workers taking greater risks is true for many ant species where it has been observed in different contexts (Hölldobler and Wilson 1990).

It is also part of nest hygiene to prevent genetically different fungal lineages from being introduced into the fungus garden, even if the foreign cultivars belong to the same symbiotic fungus species. It has been shown repeatedly that the fungus from foreign conspecific and interspecific nests often are incompatible and it was suggested that this is due to competition among genetically different fungal lineages. One should expect, therefore that the ants have evolved mechanisms to prevent such competition between cultivars within a single garden. Bot et al. (2001b) report genetic evidence that suggests the occasional occurrence of horizontal transfer of fungal cultivars not only between colonies of the same *Acromyrmex* species, but also between different sympatric *Acromyrmex* species. The ants appear to be able to assess the genetic incompatibility of different fungal strains, and eliminate the foreign fungus.

Finally, we want to address another point which is marginally related to nest hygiene. *Atta* colonies house the huge fungal body together with millions of workers and immature ants in an interconnected network of nest chambers reaching 2–6 m deep into the ground (Weber 1972a,b). This enormous biomass metabolizes and therefore produces large quantities of CO_2 which can be fatal to the ant colony if the concentration becomes too high. *Atta* workers are equipped with very sensitive CO_2 receptors on their antennae (Kleineidam and Tautz 1996). Recently, Kleineidam and Roces (2000) have measured CO_2 concentrations inside nests of *Atta vollenweideri*, and assessed the influence of nest ventilation on nest-microclimate. Because of the nest architecture, large colonies are passively ventilated. Small colonies tend to close their nest entrance during rain to protect the fungus garden from flooding. In such situations CO_2 concentrations increase rapidly, and as a consequence colony respiration rates are reduced. It is suggested that the ants' respiration remains unchanged, but the respiration of the symbiotic fungus is reduced. This, of course, negatively affects the growth rate of the fungus, and ultimately also that of the colony because the fungus is the main food source of the larvae. Thus, developing colonies are confronted with a tradeoff between minimizing risk of being flooded and drowned and providing adequate gas exchange inside their nests. The nests of mature colonies with their many vent-tubes and openings, however, provide an adequate gas exchange and microclimate (Kleineidam et al. 2001).

2.3.4 Communication and Foraging in *Atta* Colonies

The main features responsible for the tremendous ecological success of social insects are the cooperation and communication among colony members. As we have already discussed, cooperation and division of labor in leaf-cutting ants has reached a very high evolutionary grade. Accordingly, we should expect a sophisticated communication system, because the integration of the activities of thousands of individuals is only possible by means of communication. In this section we will describe some of the key features of the communication system that *Atta* colonies employ during foraging.

Leaf-cutting ants are famous for their extended foraging trails, along which they transport the harvested leaf fragments. These durable routes are very obvious to the human eye (Plates 4, 5). They lead masses of foragers to and from harvesting sites, which are either trees (mostly the canopy; e.g., *A. cephalotes*, *A. colombica*, *A. sexdens*), or large patches of savanna grass (e.g., the grass cutting *A. vollenweideri*). Early behavioral experiments by Moser and Blum (1963) indicated that the foraging trails are chemically marked with secretions from the ants' poison gland sacs. It has been suggested this trail pheromone contains a least two functional components, one volatile, which serves as a recruitment signal, and the other much less volatile, which functions as a long-lasting orientation cue. The chemical and behavioral details of the *Atta* poison gland contents still have to be elaborated, but some important aspects with respect to foraging behavior have been analyzed (Jaffe and Howse 1979; see Hölldobler and Wilson 1990).

The volatile recruitment component of some *Atta* species was the first ant trail pheromone whose chemical structure was identified (Tumlinson et al. 1971, 1972). It is methyl-4-methylpyrrole-2-carboxylate. This compound functions as a recruitment trail pheromone in at least three *Atta* species (e.g., *A. cephalotes*). In other *Atta* species (such as *A. sexdens*), 3-ethyl-2,5-dimethyl-pyrazine has been identified as the recruitment trail pheromone component (Cross et al. 1979; Evershed and Morgan 1982). In laboratory colonies, *Atta* workers readily respond to trails drawn with small amounts of these substances by following exactly these trails through all twists and turns. In fact, the effectiveness of methyl-4-methylpyrrole-2-carboxylate is quite amazing: 1 mg of this substance is sufficient to draw a potent trail three times around the earth's circumference.

The markings of the long-distance foraging routes are continuously reinforced by the foragers commuting on these trails. However, the fine tuning of the deposition of the trail pheromones and the resulting recruitment effect depends on a number of parameters, such as the quality of the food, the need of the colony's fungus for new vegetation, and others (Jaffe and Howse 1979; Roces and Hölldobler 1994). The trail pheromone and other markings also appear to affect the attractiveness of the food source and seem to stimulate harvesting activities such as cutting and carrying of leaf fragments at the harvesting site (Jaffe and Howse 1979; Bradshaw et al. 1986; Hölldobler and Wilson 1986). Not only the major trunk routes are marked with trail pheromones, but also the tree branches and twigs that are frequented by the ants. Thus, the foraging leaf-cutting ants continuously perceive the chemical trail signal. Any additional signal that mediates, for example, short-range recruitment at the foraging site, would be most effective if transmitted through a different sensory channel. Roces et al. (1993) discovered such a superposition of a mechanical signal upon the slowly fading chemical signals. In the following we present a summary of these findings.

Most harvesting by leaf-cutting ants occurs in the canopies of trees. Here, one can often observe that collectives of ants cut fragments out of particular leaves until nothing is left except a few leaf-veins, while other leaves nearby remain almost untouched (Plate 13). It appears that those leaves intensely frequented by foragers are more desirable than other leaves, perhaps because they are more tender or less loaded with secondary plant compounds, and that ants are able to summon fellow foragers to these higher quality leaves by employing special short-range recruitment signals.

In laboratory experiments, Roces et al. (1993) noticed that a number of workers cutting leaf fragments were raising and lowering their gasters, a motion pattern identical to that performed by *Atta* workers when producing stridulatory vibrations (Markl 1968). These ants possess a stridulatory organ which consists of a cuticular file on the first gastric tergite and a scraper situated on the postpetiole. By rubbing the file against the scraper the ants produce stridulatory vibrations (Fig. 5). By employing noninvasive laser-

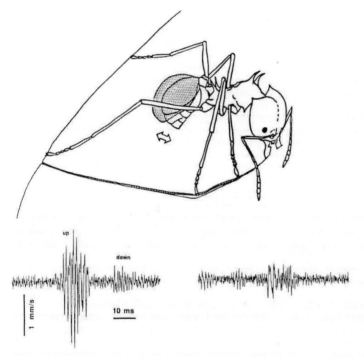

Fig. 5. Schematic illustration of an *Atta* worker cutting a leaf and simultaneously stridulating by moving the gaster up and down. Stridulation signals produced by an *Atta cephalotes* worker, recorded by laser-Doppler-Vibrometry (as velocity of the leaf's vibrations). Measurements performed 20 mm away from the cutting site. *Left* Substrate-borne vibrations transmitted mostly through the mandibles during cutting; *right* vibrations transmitted onto the substrate through the legs when the mandibles do not touch the leaf. (Roces et al. 1993)

Doppler-vibrometry, Roces and collaborators were able to record the vibrational signals on the leaf surface whenever a leaf-cutting worker stridulated. Further experiments revealed that when ants were offered leaves of different qualities, the proportions of workers that stridulated during cutting differed remarkably. Significantly more ants stridulated when tender leaves instead of thick leaves were offered. When the quality of the two kinds of leaves was enhanced because of sugar coating, almost all workers were observed to stridulate while cutting, irrespective of the differences in the mechanical properties of the material being cut. From these observations it was concluded that the production of stridulatory vibrations correlates with the quality of the leaves, and it was hypothesized that the ants that stridulate communicate leaf quality to their nearby nest mates.

Ants do not respond to the airborne components of the stridulation sound (Markl 1965), but they are highly sensitive to vibrations propagated through the substrate. A choice experiment, in which *Atta* workers on their way to the foraging site could choose between a vibrating twig or a silent one, finally proved that the stridulatory vibrations propagating along the plant act as short-range recruitment signals. However, when given a choice, more *Atta* foragers respond to recruitment pheromones than to substrate-borne stridulatory vibrations. But the effectiveness of the recruitment pheromone is significantly enhanced when it is combined with the vibrational signal. Under natural conditions, nearby workers would respond to the stridulatory vibrations transmitted through the plant material by orienting toward the source of the vibrations and subsequently joining in leaf-cutting.

The response of the ants to stridulatory signals is context-specific. Markl (1965, 1968), for example, demonstrated that workers of *Atta* species stridulate in the context of defense and alarm behavior. Roces and Hölldobler (1996) therefore argue that such rather unspecific vibrational signals used in defense and alarm subsequently acquired a function as a recruitment signal in the harvesting context (see also Hölldobler and Roces 2001).

Stridulatory signals mediate communication in an even different ecological context. In leaf-cutting ants, minim workers (the smallest worker subcaste) can be seen among foraging workers at the cutting sites, even though they are unable to cut or carry leaf fragments. Many of them do not walk back to the nest on their own, but ride ("hitchhike") on the leaf fragments being carried to the nest. It has been demonstrated that they defend the leaf carriers from attacks by parasitic phorid flies, which attempt to lay eggs on the ants' bodies (Eibl-Eibesfeldt and Eibesfeldt 1967; Feener and Moss 1990). Roces and Hölldobler (1995) provided experimental evidence that leaf carriers communicate to the hitchhikers their readiness to load up and to walk home by using plant-borne stridulatory vibrations. Minim workers nearby mount the carrier and leaf; the stridulatory vibrations produced by the carrier in this initial transport phase seem to have an arresting effect on the hitchhikers (Plate 2 below).

2.3.5 Leaf-Cutting Behavior and Load-Size Determination

A leaf-cutting forager frequently harvests leaf pieces the mass of which corresponds with her body size (e.g., Lutz 1929; Cherrett 1972a; Rudolph and Loudon 1986; Waller 1986; Wetterer 1990). This could be a result of the leaf-cutting behavior. During cutting, a worker usually anchors her hind legs on the leaf edge, and slowly pivots around the body axis, pushing the cutting mandible through the leaf tissue (Plate 2 above). In this way, the fragment size is correlated with the body size of the cutting worker (Lutz 1929; Weber 1972a,b). On the other hand, Breda and Stradling (1994) found no apparent relationship between the length of the legs of *A. cephalotes* ants and the cut curvature of the leaf fragment. The authors assume that the angle between head and thorax can be changed by the cutting ant which allows considerable flexibility. Therefore, they conclude that the fragment size "is not a simple function of the legs acting as a pivot." They also demonstrated that the leaf-cutting foragers do not directly assess fragment mass while cutting, but may use leaf toughness as an indirect measure for adjusting the size of the cut fragments.

Nevertheless, although ants can hardly cut larger pieces than their overall body size permits (unless then move along the cutting edge), it has, indeed, been shown repeatedly that they are able to change their posture in order to cut smaller leaf fragments. Before we discuss some of the arguments that aim at explaining this phenomenon, we will first briefly describe the cutting process.

During cutting, the two mandibles of *Atta* workers play different roles. While one mandible actively moves, the other remains almost fixed (cutting mandible). The steps in one bite are as follows (Fig. 6). The motile mandible is opened and anchored with its tip to the leaf tissue. The cutting mandible is not opened, but held steady. During the opening of the motile mandible, the cutting mandible is pushed against the leaf by lateral head movements. Next the motile mandible is closed, pulling the cutting mandible further against the leaf, which increases the incision. In this phase the closing mandible also moves deeper into the leaf surface, thus preparing the way for the cutting mandible. As soon as both mandibles meet, the cycle starts again.

Tautz et al. (1995) analyzed the temporal relation between mandible movements and stridulation by videotaping the cutting behavior and simultaneously recording the vibrational signals from the leaf surface with the aid of laser vibrometry. This revealed that stridulation occurred most often when the cutting mandible was moved through the plant tissue (Fig. 6). Further measurements revealed that this stridulation generated complex vibrations of the mandibles, which gave the mandibles some of the properties of a vibratome (the vibrating knife of a microtome). Indeed, when the cutting process was experimentally simulated, it turned out that the vibrating mandible clearly reduces the force fluctuations which occur when material is

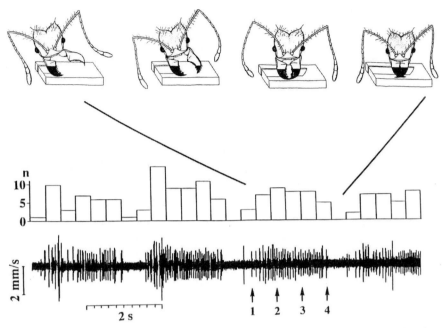

Fig. 6. *Above* Mandible and head movements during one cut into a tender leaf. *Below* Stridulation during four bites. The *histogram* shows the number of chirps counted at a bin width of 400 ms. The *trace* underneath depicts an original laser vibrometry of stridulation on a leaf measured 2 cm away from the head of the ant. The *arrows* denote the temporal occurrence of the four cutting stages shown above. (Based on Tautz et al. 1995)

being cut. Thus, stridulatory vibrations facilitate a smoother cut through tender leaf tissue (Tautz et al. 1995). It was tempting to hypothesize that this support of the cutting process was the first function of stridulation and that its employment in communication is an evolutionary derived feature. However, subsequent studies provided circumstantial evidence suggesting that the facilitating effect of the vibrations is more likely a beneficial epiphenomenon emerging from the communication process (Roces and Hölldobler 1996).

Cutting fragments out of leaves requires powerful mandible muscles (Plate 2 above). Indeed, Roces and Lighton (1995) found that the mandibular muscles in *Atta* comprise more than 50 % of their head capsule mass, or more than 25 % of the entire body mass. They also discovered that leaf-cutting is an extraordinarily energetically intense behavior. The leaf-cutting metabolic rate, which could be determined in an extremely sensitive flow-through respirometry system, is dramatically above both the standard rate and the post-cutting locomotion metabolic rate. The aerobic scope of leaf-cutting was determined to be within the same range of that of flying insects which are the most metabolically active animals. As Roces and Lighton (1995) point out, the mandibular energetics of leaf-cutting most likely play an important role in the ants' load-size selection and foraging efficiency at the individual and at the colony level.

During the past two decades, numerous papers have been published addressing the question of load-size selection in leaf-cutting ants (Plate 3). It is beyond the scope of this chapter to review in detail all the different and sometimes contradictory results. Obviously the parameters that affect load size are multifold.

As already pointed out, there is a correspondence between the size of the leaf-cutting worker and the leaf fragment size (area) to be cut. However, fragment size is not always the best parameter to determine load size (mass), because the mass of a fragment depends on leaf mass per unit surface area as well as leaf fragment size. Studies by Wetterer (1991) appear to demonstrate that in *Atta cephalotes* foragers tend not to adjust leaf-cutting behavior as a function of leaf density, but rather that workers of different sizes tend to cut leaves of different densities. These results are in agreement with those obtained by Waller (1986) who also found size-related foraging in *Atta texana*. Similar patterns of forager polymorphism and resource matching have been found in other *Atta* species (Wilson 1980b; Fowler et al. 1986b; Feener et al. 1988). However, several other independent studies found a relationship "the denser the leaf, the smaller the fragments" (e.g., Cherrett 1972a,b; Rudolph and Loudon 1986; Roces and Hölldobler 1994).

The mass of a leaf fragment being transported also affects the running speed of the carrier ant (e.g., Lighton et al. 1987; Rudolf and Loudon 1986; Wetterer 1990, 1994b; Burd 1996a,b, 2000), and both parameters (load mass and retrieval time) have important effects on the rate of vegetable material intake of the colony. However, the slower speeds of the workers carrying heavier fragments may not negatively affect intake rates, because larger loads are delivered to the colony. But extended travel time, due to heavier loads, may have other detrimental consequences. For example, the transfer of information about the food resource to the colony can be delayed and therefore the speed and intensity of recruitment can be weakened (Roces and Nuñez 1993; Roces and Hölldobler 1994).In addition, longer travel times result in increased foraging risks by leaving workers exposed longer to predators and parasites (such as the abundant phorid flies; Feener and Moss 1990; Tonhasca 1996; Tonhasca and Braganca 2000). Thus, short travel time appears to be an asset in the foraging system of leaf-cutting ants which favors load sizes that do not have too negative an effect on the running speed of leaf carriers. In any case, individual maximization models, often usefully applied to solitary foraging animals, fail to explain fragment selection by *Atta* foragers (Roces and Núñez 1993; Kacelnik 1993; Burd 1996a; for a detailed discussion of these issues see Burd 1996b). In fact, it may well be the case that small loads are rate maximizing, but at the level of the colony rather than of the individual worker. Indeed, fragment size might be influenced by the size of the worker ant, by the cutting cost, by the density (mass) of the leaf, by the need to rapidly transfer foraging information to the colony, by the distance and quality of the harvesting site, and by the "handling cost", which most likely increases with fragment size. Recently, Röschard and Roces (2002) demonstrated in the grass-cutting ant *A. vollenweideri*, where cutting cost remains the same independent of fragment size that the ants prefer load-sizes that do not hamper their movements too much during load transport. However, some handling difficulties can be circumvented by "bucket brigade" foraging, in which large leaf fragments or larger pieces of grass blades are cut by some workers and dropped to the ground for further processing or collection (Hubbell et al. 1980; Röschard and Roces 2002).

Besides such parameters as leaf tenderness or toughness (leaf mass), other traits of the plant material play a significant role in affecting the ants' preference for particular harvest material. Among the most important traits are nutrient contents and the presence and quantity of secondary plant chemicals. For instance, Howard (1988) studied the harvest selection of *Atta cephalotes* by offering the ants leaves of 49 woody plant species from a tropical deciduous forest in Costa Rica. He found that leaf protein content was positively correlated with the number of fragments cut, and that secondary chemistry and nutrient availability interact in determining the attractiveness of plant material to the ant foragers. In another study Nichols-Orians and Schultz (1990) report that young tender leaves of the tropical legume (*Inga edulis*) are more loaded with secondary chemicals and contain less nutrients than mature leaves, but the latter are three times tougher and thus harder to cut. Based on their comparative investigations of the acceptability of *I. edulis* leaves the authors suggest "that the quality of colony's habitat may indicate whether a colony will harvest more of the less suitable leaves. Colonies which are harvesting from highly suitable host plants avoid the tropical legume *I. edulis* while those in poorer habitats accept *I. edulis* but, because of leaf toughness, mostly harvest the less suitable young leaves." This in turn, suggests that harvest preferences in leaf-cutting ant colonies not only depend on particular leaf traits, but are also affected by ecosystem parameters. Thus, comparative bioassays focusing on a small set of variants often do not capture the complex multivariate picture of harvest selection.

There exists a rich and sometimes contracting literature on food plant selection in attine ants which is partially reviewed in Nichols-Orians and Schultz (1990), Nichols-Orians (1982, 1991a,b,c) and Folgarait et al. (1996; see also Chap. 10, this Vol.).

2.3.6 Communication Between the Fungus Garden and Workers in *Atta* Colonies

When two kinds of organisms live in close mutualistic symbiosis, as is the case in leaf-cutting ants and their fungus, we should expect communication between the two mutualists. The fungus may signal to its host ants its preference for particular vegetable substrates, the need for a change of diet in order to maintain nutritional diversity, or the rejection of a substrate, because it is harmful to the fungus. To date, only a few studies have examined the possibility of communication between the fungus and the host ants.

We already referred to the work by Bot et al. (2001b) that demonstrated that the ants recognize their own symbiotic fungal strain and protect it against competing strains introduced from other colonies. The authors provide circumstantial evidence "that chemical substances produced by the fungus garden may mediate recognition of alien fungi by the ants.

Selection of leaf material harvested by the leaf-cutting ants is dependent on both the physical and chemical characteristics of the plant (e.g., Cherrett 1972a,b; Littledyke and Cherrett 1978; Hubbell et al. 1983; Howard 1988; Nichols-Orians 1982, 1991a,b,c). If the plant material is loaded with secondary compounds that might be harmful to the fungus, this may cause the workers to cease harvesting of these plants. However, this reaction may not be an immediate one, i.e., it may take several hours before the foragers completely abandon this food source. Knapp et al. (1990), who have analyzed this behavior, call this "delayed rejection". However, once the colony has commenced rejecting a particular plant material, it will continue to refuse it for days or even weeks. How is the information that harvest material is unsuitable for the fungus, transmitted to the foragers?

In laboratory experiments, Ridley et al. (1996) demonstrated that the ants learn to reject plant material that contains chemicals harmful to the fungus. They provided baits containing orange peel laced with a fungicide (cycloheximide). Although the foragers initially carried the baits into the nest, eventually they stopped foraging on the bait and this rejection was maintained for many weeks. Interestingly, these test colonies also rejected orange peel not contaminated with the fungicide substance. The authors hypothesize that if the substrate causes toxic effects on the fungus, the fungus will produce a chemical signal which "acts as a negative reinforcement to the ant servicing that particular fungus garden." In a follow-up study North et al. (1999) attempted to trace the pathway of this putative fungal signal. Their results suggest that a volatile signal produced by the fungus does not affect the foragers directly, but rather that nonforager workers have to have contact with the fungus for the information regarding the fungal substrate to be transmitted through the worker force. In fact, some of their results suggest that the information is transferred from smaller fungus-garden workers to the larger forager-workers.

These are preliminary results, and the putative chemical signal (semio-chemical) produced by "stressed" fungal tissue has not been characterized behaviorally or chemically. North et al. (1999) also discuss an alternative hypothesis that proposes rejection may occur "when ants detect fungal break-down products from unhealthy or dead fungus. The workers would then asso-ciate dead fungus with 'orange flavor' and consequently reject all substrate containing orange." Indeed, it is known that leaf-cutting ants learn the odors associated with food (Roces 1990, 1994), and this is also indicated by the find-ings discussed above, where foragers that previously experienced orange peel laced with fungicide subsequently rejected noncontaminated orange peel (Ridley et al. 1996). Recently, Howard et al. (1996) have demonstrated in *A. colombica* that foragers can be conditioned to particular food sources, and studies by Roces (1990) suggested that workers recruited by scouts may be conditioned to the odor of food carried into the nest by the nestmates and this, in turn, may affect the recruits' food-selection behavior. Thus, direct and indirect conditioning to food odors most likely play a role in the harvesting process of leaf-cutting ants. The question remains as to how the fungus-gar-den workers perceive the health of the fungus and how information about the unsuitability of a particular material as fungal substrate is transmitted to the foragers. The hypothesis that this information transfer is mediated by a chem-ical stress signal emitted from the fungus is intriguing, but the supporting evidence is not yet convincing.

Color Plates

Plate 1. Areal view of Barro Colorado Island (BCI), Panama, from NW (Photo reproduced with permission from the copyright holder, C. Ziegler)

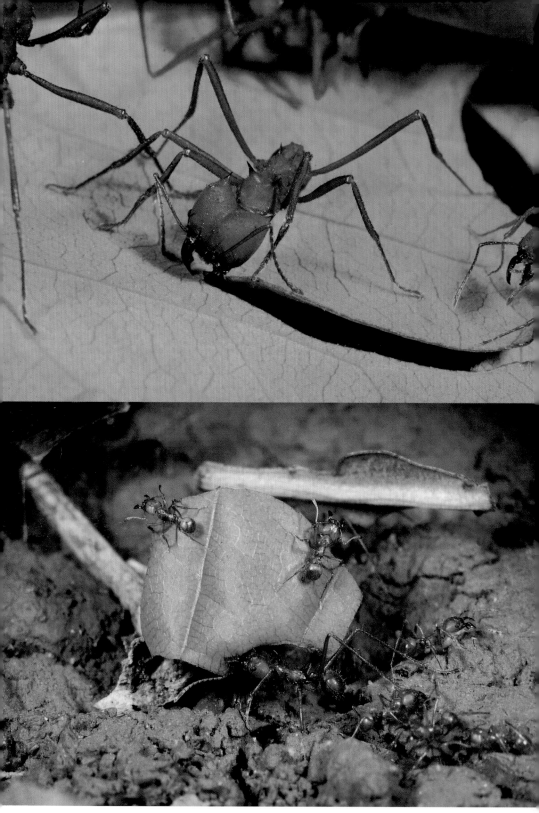

Plate 2. *Above* Forager of *Atta sexdens* cutting a fragment out of a leaf. (Photo B. Hölldobler). *Below* Forager of *Atta cephalotes* carrying a leaf fragment. Two minim workers ride on the leaf protecting the carrier ant against attacks by parasitic phorid flies. (Hölldobler and Wilson 1994)

Plate 3. Forager of *Atta colombica* lifts a freshly cut fragment. (Photo H. Herz)

Plate 4. Trail of *Atta colombica* foragers carrying leaf fragments. (Photo H. Herz)

Plate 5. Lawn on BCI showing a cleared and heavily used trunk trail of *Atta colombica*. (Photo H. Herz)

Plate 6. *Above* Section of the fungus garden of an *Atta sexdens* colony. Note the different stages of the processed leaf material (Photo B. Hölldobler). *Below* Colony founding queen of *Atta sexdens* with her worker offspring sitting on the incipient fungus garden. (Photo B. Hölldobler)

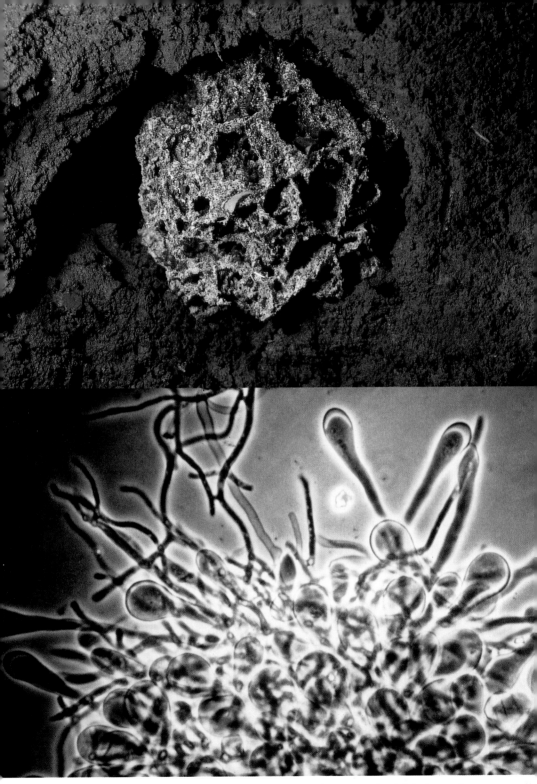

Plate 7. *Above* A single fungus garden of *Atta colombica*. Note the color change from greyish in the upper part with freshly implanted plant particles to light brown in the lower part, where the substrate is exhausted and removed as refuse (Photo H. Herz). *Below* Microscopic view of a cluster of hyphae with inflated tips, the "gongylidia" of the mutualist of *Atta sexdens*. The gongylidia are each 30–50 μm in diameter. (Photo B. Hölldobler)

Plate 8. *Above* Filamentous hyphae of the parasite *Escovopsis* overgrowing a garden of *Atta colombica* (Photo H. Herz). *Below* A queen and major workers of an *Atta colombica* colony on the trail during emigration to a new nest site. (Photo H. Herz)

Plate 9. Worker of an *Atta colombica* colony carrying a piece of the fungus garden during the emigration to a new nest site. (Photo H. Herz)

Plate 10. Refuse mound of *Atta colombica* on BCI. (Photo H. Herz)

Plate 11. Section of a refuse mound. Workers transporting exhausted fungus material climb a liana to drop the debris-particles. (Photo H. Herz)

Plate 12. Sapling of *Ochroma pyramidale* attacked by *Atta colombica* in a gap on BCI. (Photo H. Herz)

Plate 13. *Above* Leaves of *Pseudobombax septenatum* with typical signs of leaf-cutting ant herbivory (Photo A. Weigelt). *Below* Canopy of a *Pseudobombax septenatum* tree defoliated by leaf-cutting ants. (Photo R. Wirth)

Plate 14. *Above* Surface of a large nest of an active *Atta colombica* colony in the understory of BCI. Note the vegetation free "nest clearing" in contrast to the surrounding understory (Photo H. Herz). *Below* The same nest area as above ca. 4 years after the colony emigrated to a new nest site. The vegetation has started to recolonize the area. (Photo reproduced with permission from the copyright holder, C. Ziegler)

Plate 15. *Above* The same nest area as in Plate 15 ca. 6 years after the colony emigrated to a new nest site. The tree in the center of the former nest has collapsed and the vegetation reoccupied the area (Photo H. Herz). *Below* Excavated soil at the new nest site shortly after a colony movement. Note the aeration and entrance holes at the surface and the understory vegetation which is still intact. (Photo reproduced with permission from the copyright holder, C. Ziegler)

Plate 16. Artistic arrangement of the diversity of plant fragments collected by one colony of *Atta colombica* on BCI during a 1-year period. (Photo reproduced with permission from the copyright holder, C. Ziegler)

3 The Study Area – Barro Colorado Island

3.1 Location and History

The present study was performed on Barro Colorado Island (BCI) in the Republic of Panama (Plate 1; Fig. 7). Located about mid-way (9°09′N, 79°51′W) between the Atlantic and the Pacific Oceans, the island lies within an artificial lake (Lake Gatun) created by impounding the Rio Chagres during construction of the Panama Canal between 1911 and 1914. Like various other islands, BCI is high land which was isolated from the surrounding terrain as the lake filled. The island covers an area of 1564 ha. and, due to the numerous bays, a shoreline of 65 km. Located on a central plateau, the highest point on the island is 171 m above mean sea level and 140 m above the shoreline. The relief of the island appears very irregular, consisting of gentle hills traversed by gorges and steep slopes. The island is well drained by small watercourses and creeks. BCI was declared a nature reserve in 1923 shortly after its isolation and since 1946 it has been under the administration of the Smithsonian Tropical Research Institute (STRI) which operates a modern field station on the northeastern shore. Due to decades of research activities, BCI is probably the most thoroughly studied piece of tropical rainforest in the world (Leigh and Wright 1990).

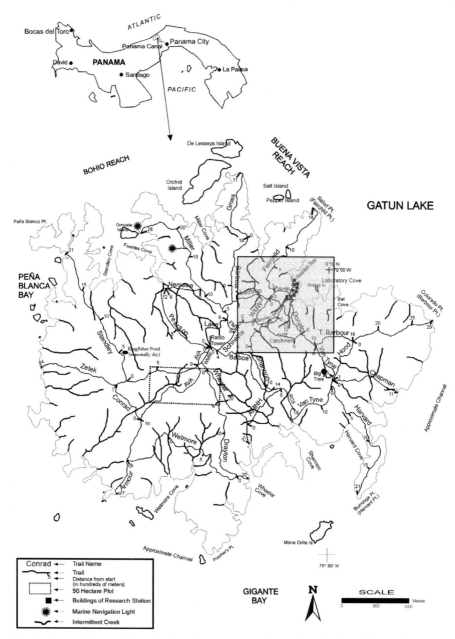

Fig. 7. Map of Barro Colorado Island and its location within the Republic of Panama. The study area is indicated by the *gray square*. (Courtesy of the Smithsonian Tropical Research Institute)

3.2 Geology and Soils

The geology of BCI is a result of the formation of the Isthmus of Panama between 2 and 3 million years ago (Marshall et al. 1979). The rocks of central Panama belong to three main types: (1) dense, impermeable volcanic rock, (2) chemically unstable stratified rock and (3) deposits of volcanic ash and silt. Like many of the large hills in the canal zone with cores consisting of intrusive or extrusive basalts (Woodring 1964), the central plateau of BCI tops an andesitic intrusion covering an area of about 3 km^2 (Leigh and Wright 1990). The rest of the island consists of sediments of terrestrial, marine and volcanic origin (Woodring 1958). The soils reflect the geology of the native rock. The intrusive rock is covered by strongly weathered soils which are relatively low in nutrients and contain accumulations of aluminum and iron oxides (so-called oxisols in the North American classification system). In the sediment areas, soils are relatively nutrient-rich with clay strata and a moderate to high base saturation (so-called alfisols; Leigh and Wright 1990; Windsor 1990).

3.3 Climatic Conditions

The climate on BCI is typical for tropical lowland regions. Monthly temperature averages vary little over the year, giving an annual mean of around 27 °C (Dietrich et al. 1982). Daily temperature variation is more substantial. Mean differences between the daily maximum and minimum amount to 9 °C. Mean annual precipitation approaches 2700 mm. Ninety percent of it occurs during the rainy season between April and December (Windsor et al. 1990; see Fig. 25B for distribution of precipitation during the present study). As a result, the long-term average of (weight based) soil moisture content at depths between 0 and 10 cm varies between 42.8 % in November and 31.4 % in March. During the rainy season, rain is most common during the early afternoon hours (between 1 and 4 p.m.), and often occurs in the form of short deluges accompanied by thunderstorms (see Windsor et al. 1990 for more details). Seasonal variations in average incident Photosynthetic Photon Flux Density (PPFD) are mainly caused by the increase of cloud cover during the rainy season. In 1994, for example, the average PPFD was 1014 E m^{-2} month^{-1} during the dry season and 789.6 E m^{-2} month^{-1} during the rainy season (data courtesy of Smithsonian Research Institution, ESP-Program). During the dry season, the climate on BCI is influenced by dry trade winds with average speeds between 6 and 9 km/h. These values are two to three times higher than the average wind speeds during the rainy season. Due to the location within Lake Gatun, the relative proximity of two oceans and the transpiration of the vegetation, relative air humidity on BCI is rather high all year-round. Long-term monthly averages (measured in a forest clearance) varied between 66.8 % RH in March and 83.3 % RH in August.

3.4 Vegetation Structure and Composition

The vegetation on BCI is classified as semideciduous tropical moist forest (Holdridge et al. 1971), with a large fraction of canopy trees losing varying quantities of leaves during the dry season from January to May (Fig. 8; Croat 1978; Leigh and Windsor 1982). Leaves are typically shed when the (weight-based) soil moisture content decreases below 40 % (Dietrich et al. 1982). The rainforest of the eastern half of the island is about 100 years old, whereas the western half of the island is predominantly covered by an older forest which has barely been disturbed during the past 200–400 years. Detailed information on the floristic history of the area is provided by Foster and Brokaw (1982). Due to these age differences, the physiognomy of the forest is not uniform. Canopy height of the old parts varies between 30 and 40 m, with maximum heights up to 50 m. In contrast, the younger parts of the forest reach average canopy heights between 20 and 30 m (maximum tree height up to 40 m; Knight 1963). The number of actual forest species was determined as 966 (Foster and Hubbell 1990b). On average, each ha of forest contains between 50 and 60 tree species with a trunk diameter larger than 20 cm (Foster and Hubbell 1990a). Thus, despite the high soil fertility, floristic diversity on BCI is lower than in the Amazonian rainforests (e.g., in the area of Manaos with 126 species/ha; Foster and Hubbell 1990a).

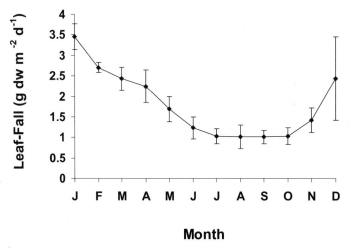

Fig. 8. Relative change of mean monthly litterfall on BCI. According to data from Leigh and Smythe (1978), collected from 1972 to 1975

The flora of BCI shows a strong resemblance to that of Central and South America. About 16% of the species are endemic for Panama and adjacent regions in Costa Rica and North Columbia, and 56% are of pan-American origin (from Mexico to South America). As a consequence, species found on the island are typical for a wide range of habitats. Some of the species [e.g., *Drypetes standleyi* (Euphorbiaceae), *Symphonia globulifera* and many other Guttiferae, *Vantanea occidentalis* (Humeriaceae)] are representatives of moist forest biotopes, whereas others [e.g., *Sterculia apetala* (Sterculiaceae), *Bombacopsis quintata* (Leguminosae), *Enterolobium cyclocarpum* (Fabaceae)], are more typical for drier woodlands or have rather large climatic amplitude [e.g., *Ceiba pentandra* (Bombacaceae), *Spondias mombin* (Anacardiaceae), *Hura crepitans* (Euphorbiaceae) and *Apeiba membranacea* (Tiliaceae)]. Detailed information on the forest floristics of BCI can be found in Croat (1978).

3.5 Fauna

BCI carries a fairly dense population of 96 species of small and medium sized mammals (Hespenheide 1994), which includes the peccary (*Tayassu tajacu*), the Agouti (*Daysyprocta punctata*), the coati (*Nasua narica*), sloths (*Bradypus variegatus* and *Choloepus hoffmannii*), the tamandua (*Tamandua mexicana*), the armadillo (*Dasypus novemcinctus*), howler monkeys (*Alouatta palliata*), capuchins (*Cebus capucinus*) and spider monkeys (*Ateles geoffroy*). Due to the small size of the island and its isolated location, big cats like the jaguar and the puma are extremely rare. The practical absence of such predators may account for relatively large populations of small mammals (Glanz 1990). Up to now, 375 bird species (Karr 1990) and 103 amphibians and reptiles (Rand and Myers 1990) have been recorded on the island. Whereas it is rather unlikely that any more unknown vertebrate species exist on this well studied island, the opposite has to be assumed for most of the insect groups. Erwin and Scott (1980), by applying insecticide fogging in Central Panama close to BCI, have extrapolated that neotropical insect fauna may run to the tens of millions of species. Therefore, individual sites such as BCI may carry tens of thousands of species. These estimates may be too high, but current knowledge about many insect groups on BCI indicates that more than 50% of the species are yet undescribed (e.g., Hespenheide 1994).

Although several myrmecological studies have been performed (e.g., Wheeler 1925; Schneirla 1933), there are no publications on the entire ant fauna of BCI. In a preliminary unpublished compilation of self-identified and literature-documented taxa, R.E. Meier (Männedorf, Switzerland, 1994) lists about 196 species. One study (Weber 1941) deals with the distribution of the entire tribe of fungus cultivating ants (Attini) on BCI. Twenty-one species and subspecies from eight genera are listed here; a unique diversity relative to the size of the island.

3.6 Research Locations

The study was performed on two spatial scales. Colony dynamics (Chap. 7) were investigated for the whole island. Studies pertaining to a single colony (Chaps. 4–6, 8–11) were conducted in a remnant of mature forest (Foster and Brokaw 1982; R.B. Foster, pers. comm.) located on the island's northeast slope, about 200 m southwest of the Smithsonian Tropical Research Institute field facility (Fig. 7, study area).

4 Species Composition of the Forest

One of the most striking features of tropical forests is their extraordinary species diversity (MacArthur 1969; Gentry 1982). The diversity of tropical trees supports a remarkable diversity of vertebrate and insect herbivores that have been shown to inflict a more intense pest pressure than in temperate settings (Landsberg and Ohmart 1989; Coley and Barone 1996). As a consequence, both discussions on plant diversity influencing (e.g., Erwin and Scott 1980) and being influenced by herbivore diversity (e.g., Janzen 1970; Connell 1971) have been of far-reaching importance in tropical ecology (for reviews see Crawley 1997; Leigh 1999).

Abundance, species composition, and foraging behavior of leaf-cutting ants is strongly dependent on the vegetation composition, i.e., the potential host plants and their diversity (Fowler 1983; Vasconcelos and Fowler 1990). For the investigation of any possible ant effects on the rainforest ecosystem it is, therefore, necessary to thoroughly assess the floristic diversity and the age distribution of trees in the surroundings of the ant colonies.

4.1 Selection and Setup of the Study Plot

The huge foraging areas of leaf-cutting ant colonies (see Chap. 9) preclude a thorough analysis of the floristic diversity of the entire foraging territory within reasonable time limits. Since mature leaf-cutting ant colonies characteristically exploit their surroundings along a few independent main foraging trails often arranged in three or four directions, a 2206-m^2 plot including one of the four main foraging trails of a representative colony was selected as a study plot representative for both the colony and the mature forest. The colony entrance hole for this principal trail was located on the eastern edge of the plot. At its northwestern edge, the plot adjoined the laboratory clearing of the research station and was, therefore, irregularly bounded in this area. The terrain sloped to the east with a mean gradient of about 30° (max.: 44°). Independently of the surface relief, the plot was subdivided into 2.5 × 2.5 m squares (Fig. 9).

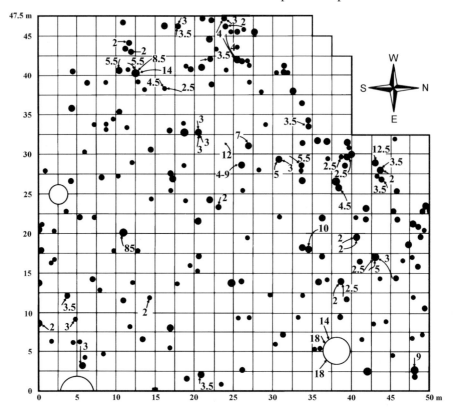

Fig. 9. The study plot with the 2.5 × 2.5 m gridlines and the mapped locations of trees and lianas

dbh classes

- 5-10 cm   • 10-20 cm
- 20-50 cm • 50-100 cm
○ >100 cm ↑ liana + cm dbh

4.2 Vegetation Inventory

In each of the 353 grid squares of the study plot, trees and shrubs with a trunk diameter larger than 5 cm at breast height (dbh) were identified and their locations mapped (Fig. 9). Lianas were also included when they relied on a tree within the respective grid square for support and their thickest stem at breast height exceeded 2 cm. Locations of Liana roots outside the grid square of the supporting tree stem were also recorded and mapped. Species were identified with the help of Croat's Flora of BCI (Croat 1978), the reference herbarium, and the staff botanists of the research station. Table. 1 lists a total of 69 tree and 22 liana species within the study plot. The species name of plant individuals with dbh values below 5 cm, but exceeding 2 m in height was not

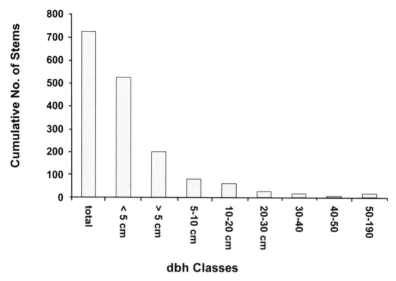

Fig. 10. Frequency of stem diameter classes (dbh) of all woody plants within the plot having heights exceeding 2 m (cumulative representation)

determined, but they were counted for each grid square as a measure for understory density. Herbs and plants having heights of less than 2 m were ignored. The size class distribution shown in Fig. 10 is based on nine trunk diameter classes out of a total of $n=722$ individuals of woody plants of at least 2 m height in the study plot. Supposing that dbh is as a rough measure for tree age, the distribution shown in Fig. 10 indicates a high recruitment rate and a stable age structure (see Schaefer 1992) of the whole tree community, although not necessarily for the different populations of species.

4.3 Abundances and Diversity of Tree Species

Floristic diversity of the study plot was calculated according to Fisher et al. (1943). Species numbers, abundances, and calculated diversity values for the various tree size (dbh) classes in the study plot are given in Table. 2. Comparing the diversity indices of the study plot with those published for BCI's 'forest dynamic plot', an area of 50 ha of mature forest, thoroughly inventoried since 1980 (Foster and Hubbell 1990a), revealed great similarities to Fisher's α for all dbh classes, indicating that the selected study plot was in fact representative for this type of mature forest. Particularly the high diversity of large trees (dbh >30) indicates that the study area was located within an old forest rather than in a successional forest area (see Foster and Brokaw 1982).

Table 1. Alphabetical list of the tree and liana species found in the 2206-m² study plot (taxonomy according to Croat 1978)

Tree species	Tree species	Liana species
Alseis blackiana	*Licania platypus*	*Arrabidaea patellifera*
Anacardium excelsum	*Luehea seemannii*	*Banisteriopsis cornifolia*
Andira inermis	*Macrocnemum glabrescens*	*Doliocarpus major*
Annona hayesii	*Maquira costaricana*	*Doliocarpus* sp.
Annona spraguei	*Miconia argentea*	*Entada monostachya*
Apeiba membranacea	*Myriocarpa yzabalensis*	*Ficus* cf. *obtusifolia*
Astrocaryon standleyanum	*Ocotea cernua*	*Ficus pertusa*
Byrsonima crassifolia	*Oenocarpus panamanus*	*Hippocratea volubilis*
Capparis frondosa	*Olmedia aspera*	*Hiraea grandifolia*
Casearia aculeata	*Phoebe mexicana*	*Hiraea* sp.
Cassipourea elliptica	*Posoqueria latifolia*	*Maripa panamensis*
Cecropia insignis	*Poulsenia armata*	*Omphalea diandra*
Celtis schippii	*Protium costaricense*	*Philodendron* sp.
Chrysophyllum panamense	*Protium panamense*	*Prionostemma aspera*
Cordia lasiocalyx	*Protium tenuifolium*	*Souroubea sympetala*
Cupania sylvatica	*Pseudobombax septenatum*	*Tetracera hydrophila*
Dendropanax arboreus	*Quararibea asterolepis*	*Tontolea richardii*
Desmopsis panamensis	*Quassia amara*	unid. 1
Didymopanax morototoni	*Randia armata*	unid. 2
Dipteryx panamensis	*Sloania ternifolia*	unid. 3
Drypetes standleyi	*Socratea durissima*	unid. 4
Faramea occidentalis	*Sorocea affinis*	unid.5 (Bignoniaceae)
Ficus citrifolia	*Spondias mombin*	
Guarea guidonia	*Swartzia simplex* var. *grandiflora*	
Guazuma ulmifolia	*Tetragastris panamensis*	
Gustavia superba	*Thevetia ahouai*	
Hasseltia floribunda	*Tovoitopsis nicaraguensis*	
Heisteria concinna	*Trattinickia aspera*	
Hirtella triandra	*Trichilia tuberculata*	
Hura crepitans	*Trophis racemosa*	
Hybanthus prunifolius	*Unonopsis pittieri*	
Inga pezizifera	*Virola sebifera*	
Inga sapindoides	*Virola surinamensis*	
Jacaranda copaia		
Lacmellea panamensis		

Table 2. Tree diversity (Fisher's α) of different size classes within the study plot (2200 m^2) and in an area of BCI's 50-ha plot (5000 m^2)

Dbh	2200-m^2 plot[a]			5000-m^2 plot[b]		
	Cumulative individuals (N)	Number of species (S)	Diversity index[c] (α)	Cumulative individuals (N)	Number of species (S)	Diversity index (α)
$\geq$ 5 cm	–	–	–	199	69	37.6
$\geq$ 10 cm	417	93	37.15	120	52	34.9
$\geq$ 20 cm	158	55	29.9	58	32	29.3
$\geq$ 30 cm	85	36	23.57	35	21	22.2

[a] Present study.
[b] From Foster and Hubbell (1990a).
[c] Diversity index $\alpha = N(1-x)/x$ (value of x estimated as $S/N = [(1-x)/x] - \ln(1-x)$, where S = total number of species and N = total number of individuals (Fisher et al. 1943).

5 Forest Light Regimes

Light availability is both a cause and an effect of forest dynamics. It is a major environmental factor limiting growth and survival of many forest species (Chazdon et al. 1996; Montgomery and Chazdon 2002) and its heterogeneity is thought to be crucial for the maintenance of species diversity in old growth forests (e.g., Denslow 1987; Canham et al. 1994). As a potential consequence of foliage loss caused by leaf-cutting ants, changes in the penetrability of the canopy for light may be expected. In order to demonstrate such effects, it is necessary to analyze the spatial and temporal light variability patterns in the forest understory as well as within the canopy. In this study, relative irradiance (RI; i.e., the fraction of photosynthetic photon flux density, PPFD, reaching a sensor somewhere in or below the canopy relative to the PPFD flux above the canopy) was chosen as a measure for light penetrability (Wirth et al. 2001b).

5.1 Seasonal and Spatial Light Patterns in the Understory

5.1.1 Measuring Irradiance

Representative measurements of total irradiance within plant stands are rather difficult to perform. Due to the extremely variable formation of light patches in time and space, the availability of the direct radiation component within a stand is extraordinarily inhomogeneous (Chazdon and Pearcy 1986, 1991). Since the goal of the present study was to identify relative changes in light penetration in certain parts of the canopy due to the activities of leaf-cutting ants rather than investigating the total irradiation supply of these areas, the radiation measurements were restricted to the diffuse component of PPFD only. This was achieved by performing the measurements either under a dense and homogeneous cloud cover or by taking measurements at dawn and dusk.

All measurements were performed with two simultaneously operated quantum sensors (LI-190SB; Li-Cor, Lincoln, NE, USA). These sensors typically record radiation incident from the entire hemisphere. Since locally distinct measurements were needed for the present study, 1.5-cm-high shades of black cardboard were mounted around the sensor working below the canopy to eliminate side lighting, but still keep the incident radiation at a measurable level. Readings at 1-m height, with the sensor and side-lighting shade in a stand of 35-m height, measured RI from a ca. 17-m radius circle at the upper canopy surface (Fig. 11). The reference sensor was mounted above the canopy at a height of about 42 m on a nearby steel tower and connected to a data logger recording data at intervals of 1 min. Within the forest stand at each grid point of the plot area, three measurements were taken with the measurement sensor at intervals of 5–8 s and also stored in a data logger (cf. Wirth et al. 2001b). This method of determining RI from measurements of diffuse radiation has the following potential sources of error: (1) any slight directly incident component in the dawn or dusk radiation would definitely be measured by the reference sensor, but not by the one in the understory. (2) Vigorous motion of the foliage, for instance due to strong winds during the dry season, might impair the reproducibility of the measurements. (3) Biases may be generated from scattered

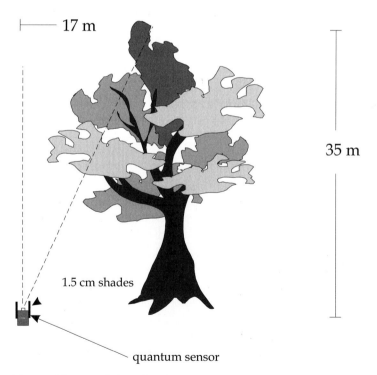

Fig. 11. View restriction of a Li-Cor quantum sensor in the understory of a forest with an assumed stand height of 35 m using a 1.5-cm cardboard shade

radiation by foliage. Several series of short-term and long-term test measurements with and without an optical filter to restrict measured wavelengths to below 490 nm (see Chap. 6.2) were executed to assure that these effects were only of minor importance in the present experimental situation.

5.1.2 Horizontal Distribution of Irradiance

Transects of RI were measured on the forest floor at all grid points of the plot area. Figure 12 shows an interpolated three-dimensional representation of the horizontal RI distribution throughout the entire plot area which was recorded at the end of the rainy season. The pattern of the frequency distribution of the RI values from Fig. 12 is slightly skewed to the left or darker side and shows that approx. 60 % of the ground surface within the plot area receives 2 % or less of the PPFD incident at the top of the canopy (Fig. 13A). The logarithmic transformation of the RI axis reveals the log-normal character in the distribution (Fig. 13B).

The methods typically used for measuring RI contain many elements of uncertainty, and the discrepancies between results cannot always be explained by differences between the various methods (Horn 1971; Alexandre 1982; Mitchell and Whitmore 1993). Nevertheless, inside mature tropical

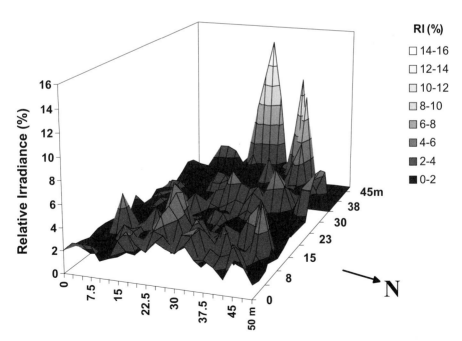

Fig. 12. RI variability at ground level within the plot area (the high values at the north-western edge were caused by an adjoining clearing)

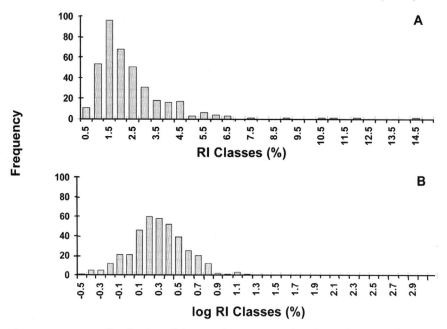

Fig. 13. Frequency distribution of the RI values measured in the understory of the plot area (*n*=384). **A** Untransformed, **B** with logarithmic values on the abscissa

forests over a wide range of latitudes the values are known to be restricted to values between 0.5 to 2.5 % of the PPFD incident at the top of the canopy (Table. 3; Leigh 1999). The fact that the RI values of the present study reach values above 2 % indicates structural differences between the semi-deciduous forest on BCI to the mostly evergreen forests from the other studies. Lee (1987) reported lower RI values for BCI. However, looking at his comparatively small sample size (*n*=35), it is hard to judge whether this discrepancy really indicates structural heterogeneity within the forest on BCI.

The RI values measured on the ground under diffuse sky conditions mainly reflect the structure and density of the foliage within the stand above. The lognormal character of their frequency distribution (Fig. 13B) can be considered typical in this respect and corresponds well with the values reported by Yoda (1974) for a West Malaysian rainforest.

5.1.3 Seasonal Aspects

Since the forest at the study site is semideciduous, radiation transmittance is expected to vary dramatically throughout the season. Therefore, RI measurements were carried out at the end of the rainy season (early December) when

Table. 3. Photosynthetic photon flux density (PPFD) as a fraction of the external radiation on the floor of various tropical forests

Location (latitude)	RI at forest floor (% PPFD) (range of variation)	Reference
La Selva[a], Costa Rica (10°26′N)	1–2	Chazdon and Field (1984)
Singapore[a] (1°20′N)	ca. 2.0	Chazdon and Field (1984)
Queensland[a], Australia (28°15′S)	0.5 (0.4–1.1)	Chazdon and Field (1984)
Pasoh-Forest[a], West Malaysia (2°59′N)	0.4	Yoda (1974)
La Selva[a], Costa Rica (10°26′N)	1.17 (0.09–3.76)	Lee (1987)
BCI[b], Panama (9°09′N)	1.49 (0.17–3.64)	Lee (1987)
La Selva[a], Costa Rica (10°26′N)	1.3–1.6 (0.6–4.1)	Rich et al. (1993)
Oahu[a], Hawaii (21°30′N)	2.4 (1.5–3.8)	Pearcy (1983)
Queensland[a], Australia (28°15′S)	0.42	Björkman and Ludlow 1972)
BCI[b], Panama (9°09′N)	2.3 (±1.3 SD)[c]	present study

[a] Evergreen rainforest.
[b] Semideciduous lowland rainforest.
[c] Based exclusively on values without influence of the clearing at the northeast edge; n=1134.

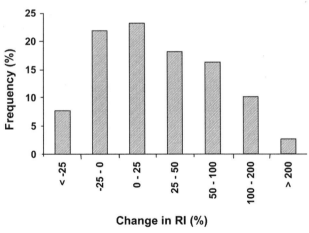

Fig. 14. Percent change in the fraction of light reaching the forest floor (RI) between wet and dry seasons measured at 375 grid points within the sample plot. (Wirth et al. 2001b, with permission from Elsevier Science)

the foliage of the stand was most abundant, at the beginning of the dry season (late January/early February) when the leaves were being shed most rapidly, and at the end of the dry season (early March) when the foliage of the stand was least abundant (see Table. 6; Wirth et al. 2001b). As expected, RI was found to differ significantly by season (Table. 4). The increase from December to

Table. 4. Seasonal variation of the relative fraction of diffuse radiation in the understory of the plot area

	Relative irradiance[a]		
	December 1993	January/February 1994	March 1994
Mean (±SD)	2.04 (±1.27)	2.22 (±1.25)	2.43 (±1.3)
n	380	376	378

ANOVA: F=8.76; df=2, 1131; p=0.0001
[a] In order to reduce the effect of the clearing at the northeast edge, only values <9 % are included.

January/February was 9.0 % and 9.1 % from January/February to March. Although RI was significantly correlated between wet and dry seasons (r=0.765, p<0.001, n=375), the relatively low correlation indicated that changes were sizeable in some regions of the forest (Fig. 14). In fact, 29 % of the grid points had increases in RI of >50 %, and at 13 % of the points, RI more than doubled. The greatest decrease in RI at a single point was 48 % and the greatest increase was 448 %.

The method used allows a relatively good description of the impact of seasonal influences on radiation transmittance through the canopy. The general similarity of the transect patterns from early December to early March indicates that the light transmittance through the canopy in the study area follows a remarkably constant pattern, despite the effects of seasonal leaf shedding. The absolute increase in the level of diffuse radiation during the dry season is consistent with the findings reported by Smith et al. (1992), but see also Lowman (1986) who observed no seasonal variability in three (two temperate and one subtropical) Australian deciduous forests (see also Wirth et al. 2001b).

5.2 Vertical Light Variability

5.2.1 Measuring Vertical Light Profiles

The general experimental setup was identical to that described in Section 5.1.1. To measure vertical transects of radiation interception through the forest canopy, some arrangement was needed to move the measurement quantum sensor vertically through the stand at various depths (cf. Wirth et al. 2001b). The principal elements of the experimental setup and its location in the study area are shown in Fig. 15. Using rope climbing techniques, a horizontal rope of 20-m length was mounted at a height of about 33 m between two trees within the plot area, which emerged slightly below the mean canopy

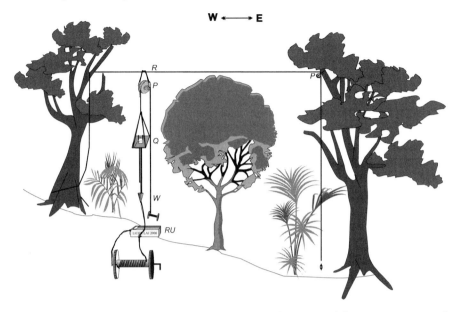

Fig. 15. Schematic representation of the experimental setup used for measurements of vertical distribution of RI within the canopy. *Q* Quantum sensor, *W* weight, *RU* recording unit, *R* rope, *P* pulley

layer. A gravity-leveled quantum sensor was connected to this horizontal rope with a pulley arrangement which allowed a person on the ground to vertically and horizontally move the sensor to any position along the main rope. Measurements were made by lowering the sensor in 2-m intervals from the upper canopy along a transect to the ground with the highest measuring point being 31.3 m and the lowest 1 m above the ground. Five vertical profiles of RI were measured along the horizontal 20-m transect. All measurements were performed at dusk (between 17:30 and 18:20 h) on 5 days in April 1994.

5.2.2 Vertical Distribution of Light

Figure 16A shows the measured RI values at various heights along five vertical transects within the study area. The means of the five profiles (Fig. 16B) represent the mean radiation attenuation through the forest stand. (Due to methodological problems, no measurements are available from the topmost 2–5 m of the canopy.) Figure 16C shows the same relationship with a logarithmic RI axis. The resulting linearity of the correlation reflects the exponential nature of the curve, which is typical for plant stands (Monsi and Saeki 1953). It is obvious that the topmost forest canopy with a thickness of about 5-m intercepts around 90 % of the incident PPFD at the top of the canopy. RI is reduced exponentially

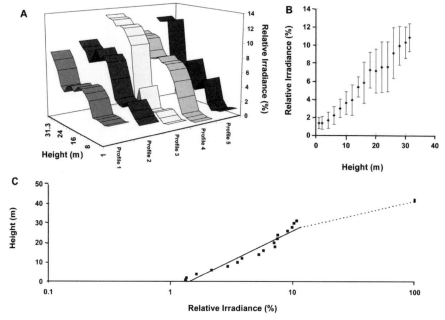

Fig. 16. A Vertical distribution of RI along five transects through the forest canopy. B Mean vertical distribution of RI calculated from the five transects in **A**; *error bars* SD. C Mean vertical distribution of RI (log-scaled) with interpolation to the canopy surface. Profiles were distributed in an E–W direction in the lower left corner of the grid area in Fig. 9

throughout the entire measuring range, thus indicating a thick layer of foliage extending from the ground to a height of about 31 m.

The measured vertical RI profiles (Fig. 16A) provide some idea of the variability in the stratified organization of the canopy foliage at various positions in the forest stand. Other than in temperate forests and particularly in monodiverse stands, where the variation in radiation attenuation follows typical curve types (Otto 1994), the situation in species-rich rainforests is less clear due to the nonuniform distribution of canopy elements with height. Nevertheless, data in Fig. 16B indicate a definite trend (which might become less pronounced if the investigation were extended to more than just one transect). Assuming a constant light extinction coefficient per unit leaf area (see below), and a homogeneous distribution of leaf area density through a stratum of the canopy in which the linear relationship holds, the foliage of this part of the BCI forest may be divided into two distinct canopy strata. In spite of the relatively small sampling area of the present study, this is quite consistent with other descriptions of the vertical structure of the forests on BCI. Foster and Brokaw (1982) reported that the mean height of the forest around the study area was about 34 m, with some crowns extending beyond this value. Various distinct stories, as reported for many evergreen tropical forests (e.g., Parker 1995) were not identified. The topmost layer in which 90 % of the incident radiation are already intercepted consists of the trees whose emergent crowns from the dense upper canopy, whereas the second stratum consists of the remaining less dense, relatively homogeneous vegetation extending down almost to the forest floor.

Although vertical gradients of light interception reflect important structural and ecological gradients within forest canopies, they are yet very poorly documented. The only known vertical distributions of RI from tall forests are from South-East Asia (Yoda 1974 in Malaysia; Torquebiau 1988 in Sumatra, and Schlensog 1997 in Borneo). Mean RI values in all canopy levels of these forests are not only lower (Yoda found 0.53 % at 2 m, 2.62 % at 14 m and 5.99 % at 20 m height; Schlensog 0.74 % at 3 m, 3.13 % at 14.5 m, and 5.90 % at 21 m height) than the present values (1.33 % at 2 m, 5.34 % at 14 m and 7.10 % at 20 m), but they also represent different stratifications of canopy foliage. Yoda, for example, divided the Malaysian forest into four distinct canopy layers instead of two, as suggested here.

6 Canopy Structure of the Forest

In order to assess herbivory rates (see Chap. 11) and also to calculate the effect of foliage removal on canopy carbon and water relations (see Chaps. 12 and 14), it is necessary to gain sufficient information on the amount of foliage present and its arrangement in space. This chapter contains such an analysis in which the vertical and horizontal distribution of foliage density (leaf area index, LAI) and foliage angles were investigated. Additionally, due to the semi-deciduous nature of the forest, the seasonal variations of LAI were also analyzed.

6.1 Theory

Foliage density and the spatial arrangement of the leaves are decisive for the exponential attenuation of light penetrating a vegetative canopy as it is described by Beer's extinction law. According to Monsi and Saeki (1953), irradiance (I) at a certain depth within a plant stand can be calculated as:

$$I = I_0 e^{-k*LAI} \tag{1}$$

where I_0 is the irradiance outside the stand and k is a vegetation-specific extinction coefficient which varies between 0 and 1 and is a function of the average foliage angle (θ). Monsi and Saeki (1953) expressed this as

$$k \cong \cos(\theta) \tag{2}$$

Thus, in a canopy of perfectly horizontal leaves k=1. Because $I/I_0 = RI$, Eq. (1) can be written as:

$$LAI = -\ln(RI)*k^{-1} \tag{3}$$

Replacing k according to Eq. (2) finally results in:

$$LAI = -\ln(RI)*\cos(\theta)^{-1} \tag{4}$$

6.2 Measuring LAI and Foliage Orientation

Since direct methods for the determination of LAI are either extremely labor-intensive, prone to error (see Kira and Yoda 1989) or destructive, various indirect methods have been developed, all of them based on the above relationship between LAI and radiation penetration (for review see Ross 1981; Norman and Campbell 1989). Gap fraction techniques, where the fraction of sky visible through the canopy is determined, became rather popular in this context. Depending on the LAI, the orientation of the foliage and the elevation of the sun above the horizon, the canopy gap fraction decreases exponentially from the top to the bottom of the stand. Thus, under certain assumptions regarding the horizontal homogeneity of the various layers of vegetation, the random azimuthal and spatial distribution of the leaves and the opacity of the leaves, LAI can be predicted. However, the assessment of canopy gap fractions by means of measuring direct radiation interception (e.g., Ross 1981; Lang 1987) is difficult in dense and tall plant stands due to the relatively small number of gaps and the occurring penumbral effects[*] (see Smith et al. 1989).

To avoid all these problems interception of diffuse light can be measured instead, since the intensity of the diffuse fraction of the light is also negatively exponentially correlated with LAI and foliage orientation. This method was used for the LAI determinations in the present study (see also Wirth et al. 2001b). A potential source of error which has to be considered if measurements of diffuse irradiance are carried out within plant stands is that the measured irradiance consists not only of light penetrating through the canopy, but also of a certain proportion of light reflected from and transmitted through the leaves which is affected exclusively by the optical properties of the leaves and is independent of the canopy structure. Since this proportion of radiation mainly consists of wavelengths above 490 nm, the error can be eliminated by equipping the used radiation sensors with optical filters which restrict the sensed radiation to wavelengths below 490 nm. To investigate the magnitude of this effect, we performed a series of RI measurements (see Chap. 5.1.1) at ground level in the grid area using a standard sensor and a sensor with the additional optical filter mounted next to each other. Statistical comparison of the obtained data with and without filter revealed no significant differences between the measured RI values (mean RI without filter: 2.73 %, with filter: 2.75 %; t=0.15, df=772, p=0.87). Thus, the fraction of scattered light was obviously low in this area, which is probably mainly due to the generally low scattering of diffuse radiation under conditions of dawn or dusk. Therefore, the

---

[*] The ring of lower irradiance around the bright central spot in pin hole images of the sun which are formed on the ground when the sun shines through a hole in the canopy of a size similar to the sun's apparent size

results of the RI measurements shown in Chapter 5 could be directly used for the calculation of LAI.

Although the values of k are not particularly sensitive to variations in leaf angle distribution, the principle weakness of many foliage light interception models is the assumption that all leaves in the canopy are inclined at the same angle (Lemeur 1973). Another frequently used but still unsatisfactory possibility is the assumption of a spherical distribution of leaf angles (Campbell 1986; Norman and Campbell 1989). In order to determine the actual leaf angle distribution at the study site more realistically, we based our estimation on direct measurements (Wirth et al. 2001b).

A steel tower near the study area was used for the measurement of the leaf angle profile. It provided access to the topmost parts of the canopy, i.e., to a height of approx. 40 m. Leaf angles were estimated at the end of the dry season at vertical intervals of 5 m, the lowest being 1 m above the ground. At each elevation, the space around the tower was divided into 12 segments with 30° angles. Within each segment the first ten leaves viewed were selected irrespective of the species and assigned to four leaf angle classes (class 1: 0°–10° (almost horizontal), class 2: 10°–45°, class 3: 45°–80°, class 4: 80°–90° (almost vertical). Positive and negative angles were not distinguished because this does not interfere with light interception. Leaves were at various distances (between 5–50 m) using a horizontally mounted telescope for the angle estimation of more remote leaves. To avoid an influence of single large tree crowns on the entire profile, the initial compass reading was increased by 10° at each height relative to that of the previous height. Thus, the resulting profile consists of samples selected along a spiral shaped trajectory extending vertically through the whole stand.

6.3 Vertical Distribution of Foliage Angles

The frequency distributions in Fig. 17 reveal the pattern well known from northern hemisphere trees (McMillen and McClendon 1979) that horizontal leaves were most common at the bottom of the forest, whereas steeply inclined blades (45°–90°, classes 3 and 4) were found mainly in the top levels of the canopy. A highly significant exponential relationship was found to exist between the mean leaf angle through the canopy and the height of the stand (Fig. 18, Wirth et al. 2001b).

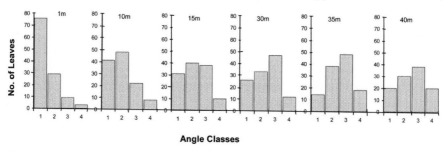

Fig. 17. Frequency distribution of four leaf angle classes at six height levels above ground (*n*=199 for 1, 30 and 35 m; *n*=120 for 10 and 15 m; *n*=109 for 40 m)

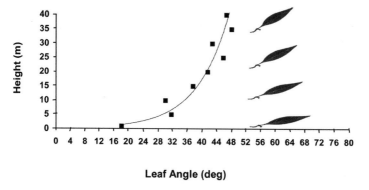

Fig. 18. Vertical distribution of mean leaf angles. Means are calculated from the medians of each angle class. The *solid line* represents the following regression equation: leaf angle=8.0252 ln(height)+16.949, r^2=0.9305, *p*=0.001, *n*=9. (Wirth et al. 2001b, with permission from Elsevier Science)

6.4 Spatial and Temporal Characteristics of LAI Distribution

Based on the estimations of RI (see Chap. 5.1) and foliage angles (see Sect. 6.3), the vertical and horizontal variation of LAI could be assessed.

6.4.1 Vertical LAI Distribution

Figure 19 shows the calculated vertical LAI distribution from the mean RI of five vertical profiles (see also Wirth et al. 2001b). It appears that more than 50% of the existing foliage was concentrated in the topmost 5 m of the canopy. Another considerable amount of foliage was located between 1 and

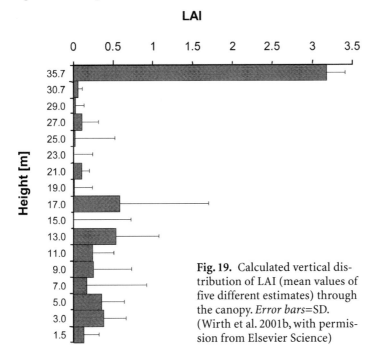

Fig. 19. Calculated vertical distribution of LAI (mean values of five different estimates) through the canopy. *Error bars*=SD. (Wirth et al. 2001b, with permission from Elsevier Science)

17 m canopy height, where many adult understory trees have their crowns. Between 17 and 30 m there was practically no considerable allocation of leaf area, because this stratum is dominated by the almost leafless trunks of trees with emergent crowns. Table 5 contains the RI values and foliage angles which were used for the calculation of these LAI data. It also gives the calculated extinction coefficients (k) for each canopy level. A mean extinction coefficient of 0.78 seems to be typical of the forest stand in the study area at the end of the dry season. Further, Table 5 compares the LAI data from Fig. 19 with LAI values calculated under the assumption of all leaves being horizontally oriented [LAI=–ln(RI)] or for an assumed spherical angle distribution [LAI=–ln(RI)*2, k=0.5]. It is obvious that the latter two methods, although frequently used in the literature (e.g., Campbell 1986; Norman and Campbell 1989), lead to considerable deviations.

The direct link between canopy light extinction and foliage orientation has been known since the seminal work of Monsi and Saeki (1953), but there has been little focus on the significance of leaf-angle distributions in diverse tropical forest stands. The observed increase in leaf angle with canopy height was found despite the variety of species sampled. This pattern, probably not unexpected in this forest, is a well documented phenomenon associated with ecophysiological optimization of water balance, light interception and carbon gain of sun and shade leaves, and has been described for temperate broadleaf

Table 5. Comparison of canopy LAI profiles calculated with three different light interception models. (Modified after Wirth et al. 2001b)

Height above ground (m)	Mean RI[a] (%)	Mean (deg)[b] foliage angle (θ)	Extinction-coefficient (k)[c]	LAI foliage angle distribution		
				Horizontal	Spherical	As measured
>40.00	100.00					
35.65	11.08	45.63	0.70	2.20	4.40	3.15
30.65	10.63	44.42	0.71	0.04	0.08	0.06
29.00	10.42	43.97	0.72	0.02	0.04	0.03
27.00	9.65	43.40	0.73	0.08	0.15	0.11
25.00	9.49	42.78	0.73	0.02	0.03	0.02
23.00	9.51	42.11	0.74	0.00	0.00	0.00
21.00	8.83	41.38	0.75	0.07	0.15	0.10
19.00	8.75	40.58	0.76	0.01	0.02	0.01
17.00	5.61	39.69	0.77	0.44	0.89	0.58
15.00	5.77	38.68	0.78	−0.03	−0.06	−0.04
13.00	3.78	37.53	0.79	0.42	0.84	0.53
11.00	3.13	36.19	0.81	0.19	0.38	0.23
9.00	2.56	34.58	0.82	0.20	0.41	0.25
7.00	2.23	32.57	0.84	0.14	0.27	0.16
5.00	1.65	29.87	0.87	0.30	0.60	0.35
3.00	1.17	25.77	0.88	0.34	0.68	0.39
1.50	1.04	20.20	0.94	0.12	0.23	0.12
Cumulative LAI				4.56	9.13	6.04

[a] Mean RI per canopy stratum calculated from five vertical RI profiles (see Fig. 16).
[b] Calculated from the data in Fig. 18.
[c] $k = \cos(\theta)$.

trees (McMillen and McClendon 1979) and tropical mangroves (Ball et al. 1988).

The importance of realistic leaf angle distributions and extinction coefficients for the indirect estimation of LAI through the inversion of a radiation transfer model seems to be obvious (Table 5). For example, the mean cumulative LAI calculated from the frequently used spherical leaf angle distribution model (Campbell 1986; Norman and Campbell 1989) is about 34% higher than the value calculated based on the measured foliage angle distribution. Similarly, the common assumption of discontinuous distributions of foliage angles seems to oversimplify the situation for complex species-rich forests (see also Lemeur 1973) and obviously fails to take into account the full range of extinction coefficient variation at different elevations within the canopy.

6.4.2 Horizontal LAI Distribution and Seasonal Variation

The ratio between cumulative LAI as measured and cumulative LAI for horizontal leaves is 1.329 (Table 5). According to Eq. (3), this factor was used as a stand-specific coefficient to transform the relative irradiance values measured at ground level at the grid points within the study plot (see Chap. 5.1.2) into LAI values. The results for the three seasonally different sets of RI measurements are shown in Fig. 20A–C. Areas with the same LAI are surrounded by isolines. Foliage density was greatest in the December wet season (Fig. 20A), most of the plot area showed an LAI of 5–6, a few hundred square meters had an LAI between 6 and 7, and only a few tens of square meters had an LAI of 7–8. The mean value for the entire plot was 5.38.

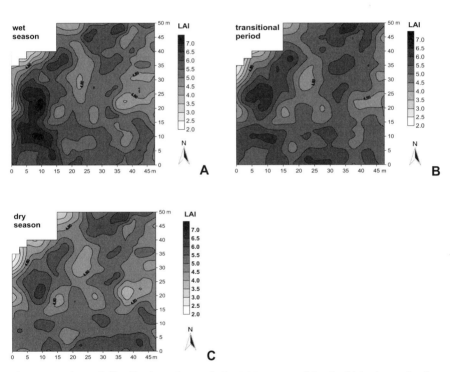

Fig. 20. Horizontal distribution of cumulative LAI at ground level within the study plot, derived from RI measurements carried out during wet season (**A**), transitional period (**B**) and dry season (**C**). According to Eq. (4), kriging was used to interpolate series of LAI measurements between grid points of the plot. The exponential interpolation method was selected since the variability of LAI was homogeneous and the grid points were regularly arranged (Isaaks and Srivastava 1989). To reduce edge effects, LAI estimated from RI values >9 % were omitted. Semivariograms were created in Variowin 2.1 (Y. Pannatier, University of Lausanne, CH) and estimated sill, range, and nugget values were used in Surfer 6.04 (Golden Software, CO, USA) to create the interpolated contour maps. (Wirth et al. 2001b, with permission from Elsevier Science)

Although the overall decrease in LAI between the wet to dry seasons was only about 5.5%, comparison of the horizontal distribution of LAI reveals considerable small scale heterogeneity (Fig. 20A–C). This is also confirmed by the relatively low correlation coefficient in spite of a high significance of the correlation between wet and dry season LAI (r=0.566, p<0.001, n=375). For instance, the large area of LAI values between 6 and 7.5 in the southwest corner of the plot steadily decreased from December through the seasons. Changes in LAI are further displayed by subtracting the values of the dry season LAI from the wet season LAI (Fig. 21). Local LAI changes at single grid points ranged from –2.3 to 2.4 m^2 m^{-2}. Although the mean LAI for the entire grid area differed significantly between seasons (Table 6), roughly 40% of the total area was not affected by these changes (considering only values >0.2 m^2 m^{-2} as changes). Foliage loss took place within an area of 1220 m^2, whereas LAI increased in an area of 130 m^2 (Fig. 22; Wirth et al. 2001b).

The measured temporal changes at the small scale found in this study strongly indicate that canopy-level means for structural parameters may not

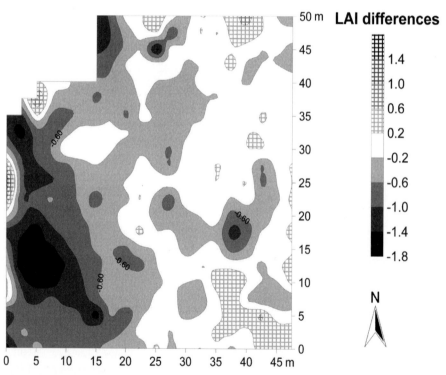

Fig. 21. Changes in LAI from wet to dry season within the 2100-m^2 study plot. Estimated LAI in December was subtracted from estimated LAI in early March at each grid point, and kriging (see Fig. 20) was used to generate contours (coloring: *filled*=decrease, *checkered*=increase, *white*=unchanged LAI). (Wirth et al. 2001b, with permission from Elsevier Science)

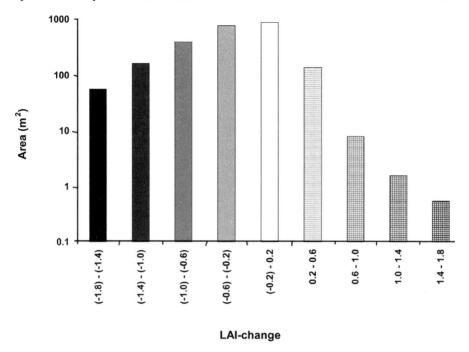

LAI-change

Fig. 22. Area of forest showing different classes of LAI change from wet to dry season. Area calculations from interpolated LAI values of Fig. 20 were conducted with Surfer 6.04 (Golden Software, CO, USA). Coloring same as in Fig. 21. (Wirth et al. 2001b, with permission from Elsevier Science)

Table 6. Seasonal estimates of mean cumulative LAI within the study plot (see also Wirth et al 2001b)

	Wet season	Transition period	Dry season
Mean LAI (±SD)[a]	5.41±0.82	5.25±0.71	5.11±0.68
n	380	376	378

ANOVA F=15.79; $p<0.001$; df=2, 1131; Tukey HSD-test: all $p<0.05$
[a] Calculated using the RI values from Table 4.

be sufficient to adequately describe functional changes in canopy structure. While large areas of the sampled plot did not show changes in total LAI – probably resulting from evergreen canopy trees that renew foliage at irregular and unpredictable intervals (Croat 1978) – we found appreciable decreases as well as local increases in LAI within the study plot. These changes in LAI corresponded to sizeable changes in the light climate reaching the canopy floor, and may have profound effects on the growth and interspecific competition of understory plants (Tilman 1988; Connell et al. 1997) and perhaps seed germination of some species (Vásquez-Yanes and Orozco-Segovia 1994).

Large tree-fall gaps are widely considered to play a prominent role for tree regeneration and maintenance of species diversity in tropical forests (Denslow 1987; Denslow and Hartshorn 1994). However, the proportion of area occupied by complete gaps is usually small, comprising between 1.4 and 7.5 % of the area in primary neotropical forests (Yavitt et al. 1995), while the area affected by seasonal changes of light in this study was much higher (29 %). Our results, therefore, imply that this dichotomy between gap and understory may be too simplistic to describe the above temporal and spatial dynamics. Thus, further studies on structural heterogeneity caused by moderate canopy dynamics in closed forests are needed.

6.5 LAI Measurements with LiCor LAI-2000

Since the above LAI estimations were all based on a single vertical transect of foliage angles (Fig. 18), a commercial LAI-measurement device (LAI-2000; Li-Cor Inc. Lincoln, NE, USA) was used for verification. (For details of the method, see Li-Cor 1989.) Measurements were taken during the rainy season at ground level along five randomly selected north–south transects within the study plot (A 90° view restrictor with the aperture facing to the north was placed on both sensors of the instrument, to minimize relief artifacts in the transmittance data.) The obtained LAI data were closely correlated with those calculated with the light interception model. Mean cumulative LAI was determined as 5.10 (±0.51 SD) by the LAI-2000 compared to 5.11 (±0.71 SD) calculated by the light interception model (paired T-test: T=0.1; df=103; $p=0.9$). With only a few exceptions (e.g., Grantz et al. 1993), a large number of studies proved the usefulness of the LAI-2000 instrument for estimating LAI. However, most of the work was done in grasslands or agricultural fields (Li-Cor 1989; Welles and Norman 1991). Studies dealing with the reliability of the instrument in more complex canopies, such as forests, are rare. Comparing estimates of the LAI-2000 with direct measurements of LAI in broadleaf forests, Chason et al. (1991) and Whitford et al. (1995) found underestimations of up to 45 % for an oak-hickory forest of moderate age and 25 % for an Australian *Eucalyptus* forest. On the other hand, Gower and Norman (1991) found good agreement with the actual value in a 28-year-old stand of *Quercus rubra*. The only application of the LAI-2000 in tropical forest stands of North Borneo resulted in leaf area index values between 4.7 and 5.1 in primary forests, and between 5.2 and 5.5 in secondary forests (Schlensog 1997), similar to the results of the present study (Table 7). Unfortunately, a verification with direct methods was not undertaken. The very good agreement between the LAI values supplied by the instrument and the data calculated with the light interception method indicates that the LAI-2000 instrument is quantitatively reliable. Since both methods are based on the inversion of a

Table 7. Mean cumulative LAI values for three different forest stands on BCI

Forest type	Old growth	Early succes-sional with dense understory	Late successional to old growth (study area)
Estimated age[a]	300–400 years	90–120 years	200–400 years
Stand height and structure[1]	30–40 m	20.0 m	ca. 34 m
LAI : MW±(SD)	5.17 (±0.55)	4.86 (±0.38)	5.14 (±0.37)
n	35	20	42

ANOVA F=3.39; df=2, 94; p=0.037
[a] Foster and Brokaw (1982)

light transmission model, the absolute values are subject to the uncertainties discussed below.

6.6 Comparison of the Study Plot with Other Parts of the BCI Forest

In order to confirm whether the above structural data are representative, further LAI measurements were conducted in a section of early successional forest and also in the old forest of the island (for details see Chap. 3). The cumulative LAI values of the two forest types were significantly different (Tukeys HSD-test: $p<0.05$; Table 7), suggesting that the young stand has not yet reached the maximum capacity of this forest type (cf. Leigh 1999). On the other hand, the results indicate that the forest community of the studied plot can be regarded as structurally similar to the old growth forest on BCI (Tukeys HSD-test: $p>0.5$) and hence confirm earlier findings based on the floristic diversity of these two forest parts (Chap. 4).

LAI values reported for other tropical forests differ so much from each other that it is difficult to assign a specific typical LAI to a particular forest formation (Table 8). A problem in such comparisons is certainly the great regional variation of structural forest complexity and the well known methodological difficulties of both indirect and direct LAI estimation techniques (see Kira and Yoda 1989; Smith et al. 1991). However, in general, it can be concluded that tropical forests on average carry about 6 ha of foliage area per hectare ground. The mean cumulative LAI of 5.4 determined within the study plot during December (maximum foliage density) appears to be rather low compared to the values typically measured elsewhere (see Leigh 1999). Since the only other available LAI estimation (LAI>7.0) for BCI was derived from litter fall measurements (Leigh and Windsor 1982) a direct comparison is difficult. Nevertheless, the data for this latter estimate were collected over a

Table 8. LAI m²m⁻² values for various tropical rainforests

Forest type/site	LAI	Reference
Montane forest, New Guinea (1977)	5.5	Edwards and Grubb
Lowland rainforest, Manaus (Brazil)	5.2	Jordan and Uhl (1978)
Lowland rainforest, Pasoh (Malaysia)	8.0	Kato et al. (1978)
Lowland rainforest, Pasoh (Malaysia)	6.9	Kira (1978)
Tropical seasonal forest, (southwest Cambodia)	7.4	Kira et al. (1969)
Semideciduous rainforest, BCI (Panama)	7.3	Leigh and Windsor (1982)
Lowland rainforest, Manaus (Brazil)	5.7	McWilliam et al. (1993)
Tropical rainforest (Ivory Coast)	3.2	Müller and Nielsen (1965)
Tropical dry forest, Guanica (Puerto Rico)	4.3	Murphy and Lugo (1986)
Terra Firme, San Carlos de Rio Negro (Venezuela)	6.4–7.5	Putz (1983)
Lowland rainforest, Manaus (Brazil)	6.1	Roberts et al. (1993)
Lowland rainforest, Sebulu (Borneo)	8.0	Yamakura et al. (1986)

period of 3 years in the same area where the study plot of the present investigation is located. One reason for this discrepancy may be that the proportion of trees and lianas which shed their leaves several times during the course of the year (cf. Croat 1978 for examples) was possibly underestimated. This would naturally lead to an overestimation of LAI. Further, it is also conceivable that our LAI values are slightly underestimated as some trees were seen to have started shedding their leaves, thereby signaling the onset of the dry season, when the RI measurements were performed in early December. On the other hand, indirect LAI determinations are actually estimations of the plant area index (PAI), because the measurement of transmitted light does not distinguish between leaves and the woody elements of the canopy. Thus, the indirectly determined LAI values should be even higher than those based on leaf shedding.

7 Colony Dynamics

The density, distribution, and size of colonies of leaf-cutting ants (*Atta* spp. and *Acromyrmex* spp.) in a particular area is of great ecological importance for their potential impact on the ecosystem through trophic, as well as non-trophic processes. Their effects as herbivores range from selective and patchy damage of individual plants up to landscape scale influence on the plant community. Their influence on the environment includes soil turnover by soil deposition from nest-building activities, nest-clearing, i.e., removal of all foliar vegetation in the nearest vicinity on and above the nest, and the accumulation of biomass in the fungus garden and in refuse mounds. The spatial and temporal distribution of *Atta* colonies varies considerably between and within their habitats. In natural forests, nest densities range from only 0.05 colonies/ha in Amazon forests (Jaffe and Vilela 1989) up to 18 colonies/ha in a forest in Pará, Brazil (Ribeiro and Woessner 1979). In cultivated areas densities can be even higher. As many as 30 colonies/ha have been reported in plantations in Brazil (Ribeiro and Woessner 1979; Jaffe 1986). Unfortunately, most authors do not specify the size of the colonies, i.e., one cannot directly draw conclusions about the magnitude of their ecological impact. Thus, the extremely high densities reported for *Acromyrmex* (up to 200 colonies/ha in a *Eucalyptus* forest, Mendes Filho in Cherrett 1989) do not necessarily imply a similarly high effect on the forest, because this genus usually has much smaller colonies than *Atta* (Cherrett 1989). Within forest ecosystems, nest densities generally tend to be higher in rather open habitats, like gaps, forest edges and clearings. In closed forests the densities decline with the maturity of the stand. Therefore, it has been proposed that high densities are indicative of disturbance (Vasconcelos and Cherrett 1995).

The queens of *Atta* belong to the longest living insects known (Keller and Genoud 1997) and their colonies are usually regarded as sessile. But long-term information on temporal changes of populations of leaf-cutting ants is extremely scarce. The only study with a duration of more than a year was conducted by Perfecto and Vandermeer (1993) who monitored the changes in an *A. cephalotes* population in Costa Rica over 2 years and reported a large increase in the population size of 22 % per year. For the tropical moist forest on BCI with its old growth and old secondary stands (see Chaps. 3, 4) three of the four Panamanian leaf-cutting ant species have been recorded: *Atta cephalotes*, *A. colombica* and *Acromyrmex octospinosus* (Rolf Meier, pers. comm., pers. observ.). Several researchers have been studying leaf-cutting ants on BCI (e.g., Lutz 1929; Weber 1941, 1969; Hodgson 1955; Porter and Bowers 1980). From their work it is known that *Atta* species have been present on the island at least since the beginning of their work, but again, little information is available on their abundance and spatial and temporal changes in distribution over that time. Very few colonies are mentioned in an early study of Weber (1969; see below). However, today, they are a conspicuous aspect of the forest on BCI. In order to assess the impact of leaf-cutting ants on the vegetation of BCI for the present study, it was, therefore, necessary to investigate the spatial and temporal dynamics of the distribution of *Atta* colonies on a long-term scale. The study investigated the entire population of the most abundant *Atta* species on BCI, *A. colombica* and lasted 5.5 years including an intense period of 2.5 years with tri-monthly to monthly surveys (see Sect. 7.1). The co-occurring leaf-cutting ant species *Atta cephalotes* and *Acromyrmex octospinosus* received less attention, because they only occurred at much lower densities than *A. colombica*.

7.1 Colony Distribution and Density

The entire island of BCI was carefully surveyed for *Atta* colonies using all available information, i.e., our own observations as well as observations of other scientists doing extensive off-trail surveys on the island. All mature nests were permanently marked. Nests were taken into account when they had grown to at least two clearly visible mounds of excavated soil or when they had two or more well established entrances. Incipient colonies[*] were excluded. Locations of all nests were mapped using a differential global positioning system (GPS Pathfinder Pro XRS, Trimble Navigation, Sunnyvale, USA) with an averaged recording of at least 100 readings for each position. A first transect-based survey of the entire island was carried out in 1993 (Wirth 1996). A detailed survey was performed in 1996. Subsequently, colony dynamics were recorded from May 1996 until October 1998 in 3-month intervals during the first year, and monthly during the rest of the investigation period. Nests were regarded as newly established when they had grown to sufficient size to be included into the survey. Newly established nests typically showed no clear signs of colony movement (see Sect. 7.2) nor did they have a clearly visible nest-clearing. Colonies were regarded as dead when there were no vis-

---

[*] Please note: the terms "nest" and "colony" are not synonymous. There can be abandoned nests without colonies.

ible ant activities at the nest site for several weeks and also no signs of colony movement (see below).

The detailed investigation of the entire island between 1996 and 1998 revealed a total of 110 *Atta* nests in 1996. Of these, 92 were *A. colombica* and 18 were *A. cephalotes* nests. Strikingly, 90 of the *A. colombica* nests were concentrated within an area of approx. 100 ha adjacent to the laboratory clearing (the so-called study area, see Figs. 7 and 23). Eight *A. cephalotes* nests were detected within the study area. The remaining ten *A. cephalotes* colonies were spread over the entire island. Because of changes due to colony movements (see below), death and the foundation of new colonies, not all of these sites were

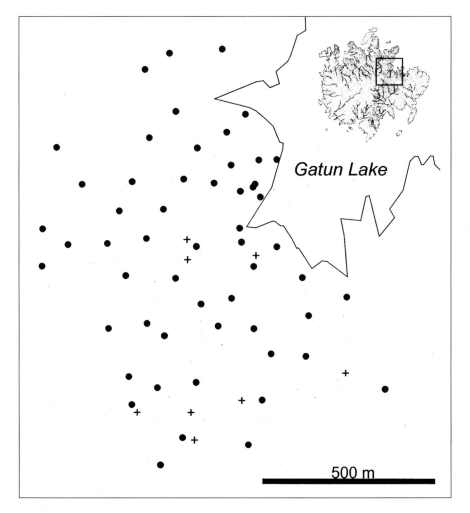

Fig. 23. Spatial distribution of *Atta colombica* colonies (*filled circles*) on BCI during August 1998. One additional *A. colombica* colony (with two nests) was located 300 m south of the other colonies (not shown). *A. cephalotes* colonies within the study area are also shown (*crosses*)

simultaneously occupied. On average, there were 52 living *A. colombica* colonies in the study area (i.e., 0.52 colonies/ha) at any one time during the investigation period plus the 8 *A. cephalotes* colonies (0.08 colonies/ha), resulting in an overall density of 0.6 *Atta* colonies/ha. Related to the entire island the density of *A. colombica* nests was 0.033 colonies/ha. The pronounced aggregation of *A. colombica* nests at the landscape level and their almost uniform distribution within the study area had already been found during the first survey in 1993 and therefore, seems to be a rather stable distribution pattern. It differs significantly [$p<0.05$, nearest neighbor analysis (Cressie 1993)] from a random distribution. If related to the entire island area, the observed nest densities of the two leaf-cutting ant species *A. colombica* and *A. cephalotes* were at the lower end of the range of reported values (e.g., Fowler et al. 1986b; Vasconcelos 1988; Cherrett 1989; but see Jaffe and Vilela 1989), while related to the study area, the values are in good agreement with literature values (but see Rockwood 1973 who found a maximum density of 2.5 *Atta* colonies/ha for an area in a Costa Rican evergreen wet forest range where the two species co-occurred). The rather uniform distribution of *A. colombica* colonies observed within the aggregation zone is well known for this species (Rockwood 1973), but has also been described for other leaf-cutting ants (see Fowler et al. 1986b for a list). Mechanisms underlying this rather distributional uniformity may be linked to resource limitation, intraspecific aggression, or high pathogen or parasite abundance in high density areas, which can either result in increased mortality rates, or in colony movements. According to the observations of the present study (see Sect. 7.2) the latter seem to play a major role in the formation of this pattern. From a historical point of view, the number of leaf-cutting ant nests on BCI appears to have increased over time. Information from Weber (1969) indicates that the abundance and density of leaf-cutting ant colonies on BCI have been much lower in former times than today. His observations dating from 1938 until 1966 revealed no more than six colonies of the genus *Atta* in total. Thus, a marked increase in the population size seems to have happened throughout the last 30 years. Reports from long-term researchers on BCI (e.g., A. Herre, STRI, pers. comm.) support this notion. The general possibility of such a quick increase in population size has been shown by Perfecto and Vandermeer (1993). They registered an annual increase of 22 % in an *A. cephalotes* population in Costa Rica. In the present study, there was no clear change in population size within the 30 months of observation, i.e., the foundation of new young colonies more or less balanced the mortality of old colonies (Table 9).

The pronounced spatial distribution difference between the two *Atta* species on BCI (*A. cephalotes* spread over the whole island, also in the parts with primary forest, at a low density; *A. colombica* only in a restricted part of the island, mainly in old secondary growths at a higher density) may indicate different habitat preferences. Similar results have been reported from primary forests and secondary growths in Brazil and Costa Rica (Rockwood 1973; Vasconcelos and Cherrett 1995). However, the question why *A. colombica* restricts

Table 9. Colony dynamics of the of *Atta colombica* population on BCI between May 1996 and October 1998; six colonies moved twice

	Number	Rate (%/year)
Colonies May 1996	51	
Colonies October 1998	54	
Colony mortality	13	9.2
Colony foundation	16	11.5
Colony movements	33	25

itself to a relatively small part of the island, although adjacent areas covered by the same type of old secondary forest (Foster and Brokaw 1982) and with similar edaphic conditions are available (see also Chap. 3) remains open. Up to now, newly founded colonies were never observed within the forest or in tree-fall gaps, but were exclusively situated within the laboratory clearing i.e., in a rather disturbed area. Similar observations were made on the mainland where higher numbers of incipient *A. colombica* colonies occurred along dirt roads or in recreational areas than within the forest (pers. observ.; C. Currie, pers. comm.). The following two possible explanations for this phenomenon are discussed, but the presently available information is not sufficient to finally decide whether these two possibilities are – alone or in combination – responsible for the remarkable colony aggregation of *A. colombica* on BCI or whether additional factors have to be considered.

1. Resource availability during the establishment phase of the colonies. During the founding phase of the colonies, the availability of high quality food sources at low cost, i.e., in close proximity to the nest may be critical for survival. Leaves which developed in the sun (as leaves of pioneer species) usually have a higher carbohydrate and nitrogen content than shade leaves (e.g., from late successional species; Coley 1983a; Chapin et al. 1987; Mole et al. 1988; Nichols-Orians 1991 c). In disturbed areas, pioneer species and sun leaves are readily available at close distance, i.e., foraging costs are markedly lower than in a closed forest (Farji-Brener 2001). In consequence, colonies founded in disturbed areas may have a higher chance of surviving the critical initial phase (Fowler et al. 1986b). Vasconcelos (1990a) used this explanation for the phenomenon that queens of *A. sexdens* typically select clearings for colony foundation.

2. Light attraction of sexuals during the nocturnal nuptial flights. It is known that freshly inseminated females start to excavate a chamber for a new colony immediately after insemination, i.e., they do not attempt to disperse or return to their mother colony (Hölldobler and Wilson 1990). Light attraction of sexuals, thus leading to insemination in the near vicinity of the light source may, therefore, cause an aggregation of nest founding sites. Buildings

and paths within the laboratory clearing at BCI are continuously illuminated at night and the attraction of sexuals of *A. colombica* to those lights was frequently observed during the period of nuptial flights. This is also supported by high numbers of *Atta* sexuals caught in the permanent insect light trap on BCI during the period of nuptial flights (N. Martinez, STRI, pers. comm.) and by observations from other (rural) locations in Panama (pers. observ., C. Currie, pers. comm.). Once grown to a sufficient size, the colonies may then colonize the adjacent forest by nest movements, which was in fact observed in four cases during the observation time (see Sect. 7.2). Thus, consecutive movements away from the clearing into the forest would explain the observed distribution of mature colonies within the area.

7.2 Colony Movements

Typically, there are three reasons for the changes in the distribution pattern of *Atta* colonies: mortality, foundations (for definitions, see Sect. 7.1), and colony movements. During the observation period on BCI, by far the most changes in colony locations resulted from movements of entire colonies. Normally, colony movements can be observed directly. Bidirectional transport of cut plant fragments to the old *and* to a new nest on the same foraging trail, transport of brood, workers, and fragments of the fungus garden from the old to the new nest site, and soil excavation at the new nest site typically indicate an ongoing colony movement. Movements which have already been completed in the near past can easily be detected by the fact that one of the typically long lasting cleared trunk trails of an apparently ceased colony leads to a newly excavated nest. Between May 1996 and October 1998, 33 movements of entire colonies of *A. colombica* could be observed. Assuming the presence of 52 living colonies (see Sect. 7.1), this indicates a moving rate of approx. 25 % per year. The average distance between the old abandoned site and the new site was 112.5 m with values ranging from 33 to 258 m. For a detailed description of the colony movements on BCI, see Herz (2001; Plates 8 below, 9). So far, reports on colony movements of *A. colombica* (Rockwood 1973; Porter and Bowers 1980), and other leaf-cutting ant species (Fowler 1981) suggested that these would be rather rare events. Therefore, the observed colony relocation frequency of 25 % of the whole population per year was unexpectedly high. The high frequency of colony movements has several important ecological implications: (1) soil turnover by excavation of new nests (Perfecto and Vandermeer 1993), (2) creation of 'gaps' in the understory through new nest clearings (Farji-Brener and Illes 2000), (3) abandonment of formerly cleared sites and its effects on the succession of understory vegetation (Garrettson et al. 1998), (4) changes in the herbivorous pressure on individual plants, and (5) accumulation of biomass and nutrients at the new location (Haines 1978).

These consequences for the forest are discussed in Chapter 15. Causes for the movements still need to be resolved. So far, there is no evidence that microclimatic stress (Hölldobler and Wilson 1990) does play a role (pers. observ.). Whether aggressions between neighboring colonies (Vilela and Howse 1986; Salzemann and Jaffe 1990a,b; Whitehouse and Jaffe 1996) or potential infections of the fungus garden (some evidence in Currie et al. 1999a; see also Herz 2001) play a role in this context is still under investigation.

7.3 Nest Size

Typically, the size of a leaf-cutting ant nest correlates with the age of the colony (Bitancourt 1941; Jonkman 1980b; Hernández et al. 1999). This allows us to draw conclusions on population dynamics from nest size data for a particular species at a given location. Unfortunately, no such information is available in the present literature. In the present study, nest size was defined as the area of the nest clearing as determined by measuring two perpendicular maximal and minimal diameters across the clearing and calculating the area as an ellipsoid. Observed nest sizes of *A. colombica* ranged from 0 (i.e., the rare case of nests without clearing and soil deposition where the size could not be determined) to 95 m². Mean nest size was 22 m² and the distribution of size classes was skewed towards smaller nests, but did not show a clear exponential pattern (Fig. 24). This suggests that there is recruitment in the population,

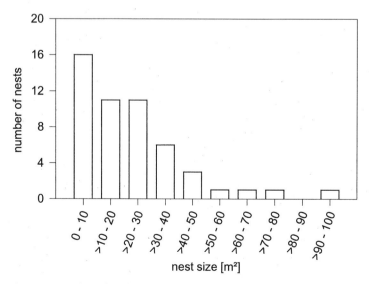

Fig. 24. Size-class distribution of *A. colombica* nests on BCI. Nests of relocated colonies were excluded (*n*=48)

but mortality in young colonies appears to be lower than would be expected for an explicit exponentiality of the age structure. On the other hand, it has to be stated that the size-class distribution is probably blurred by the high movement rate of colonies, because nest sizes of the same colony were usually lower after the movement than before. Although only established or newly founded colonies were considered for this analysis, the occurrence of movements shortly before the observation period cannot be excluded and may affect the results. The average nest size of *A. colombica* on BCI is relatively small compared to other *Atta* species. *Atta cephalotes* nests in La Selva, Costa Rica, have an average size of 30 m² (calculated from Perfecto and Vandermeer 1993) similar to the grass-cutting ant *A. vollenweideri* in Paraguay (34.6 m²; Jonkman 1980a). *Atta* nests larger than 100 m² (Cherrett 1989) were never detected on BCI. The smaller nest sizes in *A. colombica* may be due to a shorter life span of the colony, resulting in smaller worker population size in adult nests or a shorter apparent nest lifetime due to the high movement rate.

7.4 Nest Trees

Forty nest sites (42 % of all nests) of *A. colombica* were located at the base of a large tree, the 'nest-tree' (Plate 14 above). Interestingly, almost half (18) of these were fig trees (*Ficus* spp.). In a tree census in the same area from 1975, Thorington et al. (1996) found that out of a total of 856 trees (with dbh>19 cm) only 3.3 % were figs. Thus, figs are nest trees of *Atta* colonies in a significantly higher proportion than their occurrence in the tree community ($p<0.001$, $\chi^2=220.14$, df=1). The observed pattern in the present study is probably a result of the high colony movement frequency of the studied population of *A. colombica*. Typically, excavations for a new nest of a moving colony are started at the base of a tree trunk because penetration into the ground is probably facilitated here by various little breaks in the soil surface (pers. observ.). The fact that a high percentage of *A. colombica* nests was distinctly located at the base of a large tree may not be generalized for other species of the genus. Vasconcelos (1990a), for example, found colonies of *A. sexdens* more often within tree-fall gaps than in the adjacent forest. The observed remarkable association of the ants with fig trees may also be related to the high quality substrate this plant species is providing for the ants. While attached fig leaves are rarely cut due to extravasating milk sap, shed stipules and leaves are readily cut and picked up from the ground (see Chap. 8.1). There may also be an unknown advantage for the cultivation of the fungus garden in the vicinity of the roots of a fig tree (Charles Handley, STRI, pers. comm.). Whether fig trees are actively selected by the ants for these potential benefits or whether the high association rate simply reflects a higher survival rate of the young, recently moved colonies due to a better resource supply remains open.

8 Harvest Dynamics

To assess the impact of a herbivore on the vegetation it is essential to quantify the proportion of plant material which is removed from the standing crop by this particular species. The amount of harvested biomass carried into nests of leaf-cutting ants has been estimated by several authors (Hodgson 1955; Cherrett 1968; Blanton and Ewel 1985; Haines 1978). Only one publication (Lugo et al. 1973) quantifies the biomass flow caused by the activity of the ants and its influence on the energy flow in a forest. However, most of these studies cover only a short time and, thus, offer little opportunity for extrapolation to seasonally influenced habitats. In order to get this necessary long-term information on the quantitative and qualitative aspects of the harvesting activities of leaf-cutting ants and their impact on the vegetation, the foraging activities of 25 colonies of *Atta colombica* in the study area were monitored during 1993–1994 and 1997–1998. Two of these colonies were monitored in particular detail, making it possible to quantify the amount of foliage harvest per day, per season, and per year and to determine all plant species harvested throughout this time. Further, with respect to the unexplained variability of the circadian activity between leaf-cutting ant colonies (e.g., Lewis et al. 1974b; Pilati and Quiran 1996), information about the daily foraging cycle and its causes was collected (see also Wirth et al. 1997).

8.1 Annual Pattern of Foraging Activity

8.1.1 Measuring Foraging Activity

Detailed (i.e., weekly) foraging activity counts were performed at two mature colonies of *Atta colombica* (colonies I and II); random 5-min counts on different days throughout the year were taken from 49 other colonies (III–LI), 23 of which (III–XXV) were observed for 24 h each. Colony I was monitored a total of 45 census days between mid-June 1993 and late May 1994, whereas foraging activity of colony II was measured on 15 different census days between

February and May 1994. During census days, 24-h measurements of foraging rates were performed by hourly counts during the day and at 2–3 h intervals at night since nocturnal activity was minimal. Counts were done with a hand-held counter and a stop watch; laden ants were counted returning to the nest for 2 min (or only 1 min when activity was higher than 50 fragments/min) at a fixed point near the entrance holes of all foraging trails. At colonies I–II, the harvested plant material was divided into resource categories (i.e., green leaves, fruit parts, flower parts, stipules, litter or miscellaneous) which were recorded separately; at colonies III–XXV only the total number of fragments/min was counted. Hourly inputs were subsequently estimated from these data. The total input of plant parts per colony was estimated by summing up the calculated hourly inputs from all foraging trails. A "gentle" sampling method was applied by collecting ants together with their loads using a rechargeable vacuum cleaner ("Dustbuster Plus", Black & Decker, Inc., Towson MD, USA) and releasing them after they dropped the carried fragments. The fragments were then categorized and separated by species and type of plant part (leaves, fruits, flowers, stipules, litter, or miscellaneous). Subsequently, fragment area and dry weight (dw) of all pieces was determined. The data of colony I originate from samples collected during the peak phase of daily harvesting activity (see Sect. 8.3). A detailed description of sampling methods, species identification, as well as determination of fragment area and dry weight, is given in Wirth et al. (1997).

8.1.2 Plant Biomass Harvest

The collected data provide a comprehensive view of the foraging activity pattern over a 1-year period. Harvesting varied considerably from one observation day to the next, ranging from 9770 to 374,200 collected fragments per day for colony I and 6015–94,785 fragments per day for colony II. However, plotting the data as moving averages over four census days revealed distinct trends, showing that the annual foraging was strongly related to rainfall patterns (Fig. 25). Total annual harvest of plant biomass of 378 kg dw collected by colony I did not exclusively consist of green leaf material (Plate 4), but contained also a substantial fraction of nongreen material (Plate 16). Assuming a dry season of 4 and a rainy season of 8 months, it can be stated that green leaves were preferably collected during the wet season, while nongreen plant material (fruit parts, stipules and flower parts) was more prevalent in the dry season. The latter plant parts typically have a higher water content and nutritional value (e.g., Dixon 1966), and a lower content of repellent or toxic secondary compounds (e.g., Feeny 1976) than green leaves, and may be preferable as substrate for the fungus. Further, they need not be cut in the tree crowns, but just be collected from the forest floor after being shed by the trees. The increased availability of these resource types is due to the fact that many

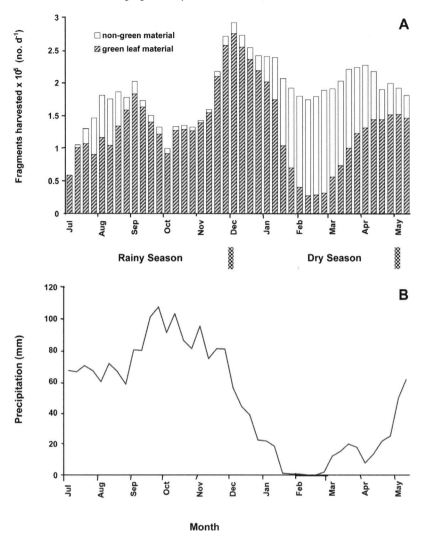

Fig. 25. Annual pattern of daily biomass harvested by colony I (**A**) and weekly precipitation totals in the study area on BCI (**B**), June 1993–June 1994. Both trends expressed as moving average over 4 days of observation. (rainfall data courtesy of Smithsonian Tropical Research Institute, Environmental Science Program). *Ticks* on the *abscissa* represent the beginnings of the labeled months. (Wirth et al. 1997)

forest plants reproduce during the dry season (Croat 1978). Thus, estimated annual total dry weight of this nongreen material accounted for almost one third (111 kg dw) of all plant material carried into the nest. The fraction of nongreen food items estimated for 49 colonies was of similar magnitude, averaging 23.9±22.9 % SD of the annual harvest, and hence, reflecting site-specific variation of resource availability.

Mean daily flow of total plant biomass into colony I was about 1 kg dw on average. Total daily flow of biomass significantly increased by 30% (t=2.78, df=43, p=0.010) from 934 g during the wet season (n=28 day, 1 day/week) to 1230 g during the dry season (n=17 days; 1 day/week; Table 10; see also Wirth et al. 1997). The general validity of this seasonal pattern was confirmed by 15 counts taken at colony II. Although the absolute harvesting rates differ due to the colony size, the seasonal pattern of lower leaf harvests during the dry season and a resumption of leaf-cutting activity at the onset of the rainy season is the same for both colonies (Fig. 26). Total yearly input of green leaf material collected by colony I amounted to 3855 m² of foliage area, which is equivalent to approx. 10 m²/day. Total area of foliage collected by colony II was 1706 m²/year, i.e., roughly half of that of colony I (Table 11). The harvesting activity of colonies I and II was within the range of annual total biomass intake (i.e., green and nongreen material) of 49 other *A. colombica* colonies on BCI (mean value 273±161 SD kg/year with single rates between 85 and 470 kg/year; see also Sect. 8.4.2 below), corresponding to a mean harvested leaf area of approx. 2700 m²/year (single values between 835 to 4550 m²/year) with the 370 kg/year value of colony I certainly being in the top range. Generally, the colony harvest rates of the present study appear to be higher than most published values. Evaluating the colony input data of 17 studies on *Atta* presented by Fowler et al. (1990) reveals a mean value of 140.35 kg/year (see also Wirth et al. 1997), but shows also a considerable variability between the different determinations (SD=±226.5/year, n=17). Fowler et al. (1990) discuss the following three major causes: (1) many of the data were collected during short-term studies (e.g., Hodgson 1955; Cherrett 1968; Lugo et al. 1973; Blanton and Ewel 1985) and, therefore, cannot simply be extrapolated to give total annual harvests, (2) the comparability of the data is questionable as in several

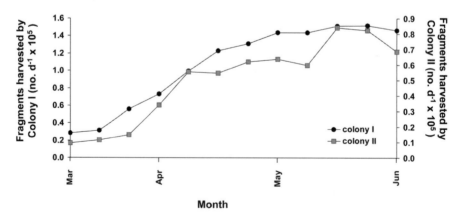

Fig. 26. Seasonal pattern of leaf biomass harvested by colonies I and II from February to June 1994. Trends expressed as moving average over 4 days of observation. *Ticks* on the *abscissa* represent the beginnings of the labeled months

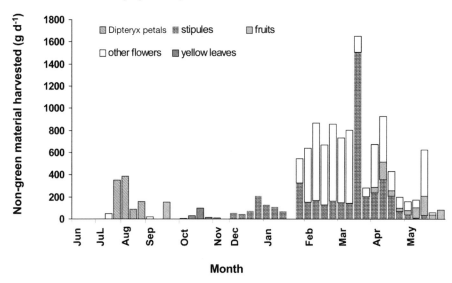

Fig. 27. Annual pattern of nongreen plant material collected by ants of colony I. *Ticks* on the *abscissa* represent the beginnings of the labeled months. (Wirth et al. 1997)

Table 10. Average daily inputs ($\pm$SE) of plant material harvested by colony I on a seasonal and annual basis ($n=45$, $n_{dry\ season}=17$, $n_{wet\ season}=28$)

	Green leaves	Nongreen material	Total biomass
Average daily inputs (g dw)			
Dry season	619$\pm$105	612$\pm$79	
Wet season	789$\pm$84	145$\pm$35	
Annual			1064$\pm$64

Table 11. Total leaf area and daily harvest rates ($\pm$SE) expressed as number of fragments and foliage area of green leaves harvested by colonies I and II on a seasonal basis (colony I: $n=45$, $n_{dry\ season}=17$, $n_{wet\ season}=28$; colony II: n=15, $n_{dry\ season}=10$, $n_{wet\ season}=5$)

	Colony I		Colony II	
Season	Dry	Wet	Dry	Wet
Total leaf area (m²)	1115	2740	324	1382
Mean daily inputs:				
No. of fragments $\times 10^5$	1.13$\pm$0.19	1.45$\pm$0.15	0.33$\pm$0.1	0.73$\pm$0.16
Leaf area (m²)	8.96$\pm$1.53	11.42$\pm$1.21	2.6$\pm$0.82	5.76$\pm$1.07

cases no information is given on the nest size (e.g., Hodgson 1955; Emmel 1967), and (3) different methods of data acquisition were used to quantify the harvests, each of them with certain drawbacks. From the experiences in this study, an additional shortcoming of many published quantifications of leaf-cutting ant harvest can be added (see Wirth et al. 1997): although the fact that sizeable amounts of the harvest consist of nongreen plant material has been described (Cherrett 1968; Lugo et al. 1973; Rockwood 1975), it was not suffi-ciently considered in annual foraging estimates. Since the different resource categories were separately sampled during the present study, a more detailed view of the role of nongreen resources during the course of the year was pos-sible. As can be seen in Fig. 27, most of the nongreen material was collected during the dry period from January to early April. Obviously, intake of non-green material was closely related to phenological characteristics of the prin-cipal forage plants. At the begin of the dry season, for instance, when many tree species started to complete their reproductive cycles, floral parts (e.g., petals of *Anacardium excelsium* and *Dipteryx panamensis*) were the domi-nant plant parts collected, and were taken in large quantities. Also, the stipules of two abundant *Ficus* trees, *Ficus yoponensis* and *Ficus obtusifolia*, were intensely collected at that time. During the latter part of the dry season and the transition to the wet season (April–June), the quantities of fruits and fruit parts (primarily from *Ficus obtusifolia*, *Ficus yoponensis*, and *Miconia argen-tea*) increased. Although flower parts were the most frequent nongreen mate-rial collected in the second half of the dry season, quantities were much smaller than in the first half and reflected reduced availability. There is evi-dence that the amount of harvested nongreen material may vary substantially between sites and phenological state of the respective vegetation. In contrast to the approx. 25 % of nongreen plant material collected by the colonies on BCI, Shepherd (1985) found that in a moist forest in Colombia fruits and flower parts accounted for only 5 % of the total input.

The data presented above reveal a considerable variability of the daily bio-mass harvest. However, relating the number of observation days to the relative coefficient of variation as a measure of harvest variability (Fig. 28) shows that the sample size was adequate. The comparison between census days with and without rainfall again illustrates the role of precipitation as a major source of harvest variation. Considering only days without rain, 8 days were necessary for a fairly constant value of variability. In contrast, 20 were needed if all days (i.e., days with and without rain) were taken into account. Besides intensity and timing of rainfall, there was also some inherent variation in daily total harvests, which has to be attributed to other factors such as resource avail-ability and distance as well as restraints of foraging activity caused by inter-specific competition as discussed below.

Observation of colonies I and II revealed rather similar seasonal activity patterns, indicating that this harvesting behavior may be generalized. Unfor-tunately, it is difficult to compare the patterns of these two colonies with the

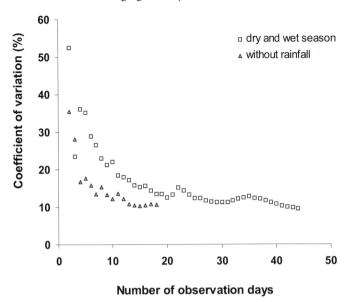

Fig. 28. Relative coefficient of variation of daily leaf harvests in colony I as a function of the number of observation days during the study year. Forty-four days from the dry and wet season were considered (*open squares*), and 17 wet season days without rainfall (*filled triangles*)

other 49 colonies mentioned above: in these, biomass harvest rates were estimated during 1997/1998, when the annual climate pattern was strongly affected by the El Niño phenomenon (very little rain). Further, data were derived by refuse-dumping activity counts (see Sect. 8.4.2). Due to delayed passage of material through the nest ("buffer effect"), this method does not allow direct conclusions about changes of foraging intensity within months. To explain the lower overall harvest rates during the wet season various reasons need to be considered: Hubbell et al. (1984) suggested that the reduced preference of *Atta cephalotes* for green leaves in the peak of Costa Rica's wet season was due to a seasonal increase in chemical repellent of leaf extracts. Another reason, which has been discussed for several primary consumers on BCI, is that their populations are controlled by the scarcity of young leaves at the height of the rainy season (Leigh and Windsor 1982). The present results indicate that the low harvesting rates during the rainy season are rather the result of the frequent reductions of foraging activity caused by rain events than of a lack of suitable leaves (*Atta* species cut considerable amounts of old leaves at the end of the wet season; cf. Morini et al. 1993; pers. observ.). Although the total input of plant material to the nest is lower during the rainy season, the harvest minimum for green leaf material occurs during the dry season. However, this seasonal minimum in leaf harvest is compensated by the increased harvest of other more abundant resources such as flower parts, fruits and *Ficus* stipules. This shift is obviously linked to the phenology of the dominant plant species within the foraging territory (Croat 1978; Foster 1982; and unpublished data of the Environmental Science Program, STRI). Strong seasonality of foraging with compensatory periodical shifts in the usage of

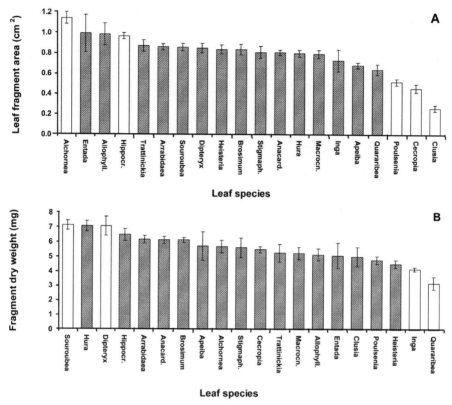

Fig. 29. Variability of mean surface area (**A**) and dry weight (**B**) of green leaf fragments (±SE) harvested by ants in colony I for 20 dominant green leaf sources. *Blank bars* represent species which were significantly different (ANOVA: fragment area: F=9.23; df=19, 117; $p<0.001$; dw: F=3.57; df=19, 117; $p<0.001$; and subsequent Tukey multiple comparisons) from most other species (*shaded bars*). The following species are shown with number of samples analyzed per species given in parentheses (total number of samples analyzed: 137, consisting of a total number of 10,224 fragments): *Alchornea costaricensis* (*n*=4), *Allophyllus psilospermus* (*n*=3), *Anacardium excelsum* (*n*=22), *Apeiba membranacea* (*n*=4), *Arrabidaea sp.* (*n*=9), *Brosimum alicastrum* (*n*=3), *Cecropia insignis* (*n*=7), *Clusia uvitana* (*n*=5), *Dipteryx panamensis* (*n*=8), *Entada monostachya* (*n*=4), *Heisteria concinna* (*n*=4), *Hippocratea volubilis* (*n*=9), *Hura crepitans* (*n*=4), *Inga pezizifera* (*n*=7), *Macrocnemum glabrescens* (*n*=16), *Poulsenia armata* (*n*=7), *Quararibea asterolepis* (*n*=5), *Souroubea sympetala* (*n*=5), *Stigmaphyllon hypargyreum* (*n*=6), *Trattinickia aspera* (*n*=5). (See also Wirth et al. 1997)

different resource types has also been described for *A. cephalotes* in the semi-deciduous forest of Guanacaste Province, Costa Rica (Rockwood 1975, 1976). Seasonal shifts in harvest may also be consistent with the predictions of the optimal foraging theory (reviewed by Pyke et al. 1977), which is also supported by the results of Roces and Lighton (1995), who found that during the cutting, *Atta* ants require many times their basal energy needs. Consequently, it is more cost-effective to gather energy-rich items from the forest floor than to engage in energy intensive leaf-cutting and tree climbing (see also Rock-wood 1975). Data on the spatiotemporal adjustment of foraging to the pattern of resource distribution, which will be discussed in Chapter 9, point into the same direction. In conclusion, it can be stated that two factors are mainly decisive for the observed seasonal pattern of foraging activities: (1) heavy rainfalls during the wet season which physically impede leaf-cutting ants, leading to a curtailment of the time available for foraging and (2) seasonal changes in the availability of green and nongreen plant material, i.e., flowering of most species and leaf shedding of the deciduous species during the dry season, as well as the start of leaf flushing at the beginning of the wet season (March) reaching a maximum in May and June (see also Fig. 25).

8.2 Size and Weight of Harvested Fragments

Since foraging activity counts can be used to be convert to biomass intake, the average size and weight per leaf fragment are of practical importance. Random samples from 49 colonies in the study area revealed an average fragment area and fragment weight of 0.86 cm^2 ($\pm$0.17 SD) and 6.51 mg ($\pm$1.91 SD), respectively (Table 12). In colony I, based on 679 samples containing a total of 17,220 leaf fragments, the colony-specific average area of leaf fragments carried by individual ants was 0.79 cm^2 ($\pm$0.24 SD), while the average fragment dry weight was 5.51 mg ($\pm$2.5 SD) (Wirth et al. 1997). The interspecific variability of the fragment size and weight for 20 predominant leaf species is shown in Fig. 29. Only a few species exhibited major differences from the above means. Succulent leaves (e.g., *Clusia uvitana*), sclerenchymatous leaves such as those of *Poulsenia armata* or leaves with a pubescent lower surface (e.g., *Cecropia obtusifolia*) were cut into relatively small pieces (<0.5 cm^2), while fragments cut from thin and glabrous leaves (e.g., *Alchornea costaricensis*) were often larger than 1 cm^2. Fragments carried by a foraging ant are more variable in surface area than in weight (Fig. 30). While the dry weight of the

Table 12. Parameters used to estimate the consumption rate of the *A. colombica* population

	Mean	$\pm$SD	n
Number of colonies	47.6	2.1	12
Monthly average of refuse loads (number min^{-1})[a]	5051	413	12
Fraction of leaf fragments from total harvest (%)	74.0	23.7	136
Average fragment weight (mg)	6.62	1.86	136
Average leaf fragment area (cm^2)	0.863	0.171	

[a] See Sect. 8.4.2 below

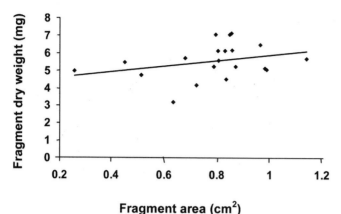

Fig. 30. Relationship between dry weight and surface area of green leaf fragments collected by colony I for the 20 dominant species listed in Fig. 29 (r=0.315, p=0.175, n=20)

analyzed fragments varied only by a factor of 2.2, fragment area fluctuated by a factor of 4.5, indicating that the ants select their loads on the basis of fragment weight rather than size as suggested by Rudolph and Loudon (1986) and Cherrett (1972b). There was no significant correlation between these two parameters (r=0.315, p<0.175, n=20). The fact that fragment weight seems to be the main criteria for load size selection can also be inferred from results published by Breda and Stradling (1994) showing that leaf-cutting ants base the size of the cut mainly on the thickness of the material. Besides leaf thickness, the observed variability in load weight can depend on various other leaf parameters such as toughness (Roces and Hölldobler 1994), pubescence (Howard 1988), or water content (Bowers and Porter 1981).

Regarding the load sizes of nongreen plant material, a considerable variation could be observed. Thus, except for the rather lightweight flowers, no significant differences between the average dry weight of four classes of nongreen plant parts could be detected (Fig. 31; Wirth et al. 1997). This high variability reflects the observation that most of the pieces were collected whole rather than cut. Fragments such as *Ficus* stipules weighed about one third (8.5 mg ±2.77 SD) more than the average leaf fragment and almost twice as much as an average flower part (4.83 mg ±2.52 SD). Thus, depending on the kind of available resources, the biomass input into a leaf-cutting ant nest can vary considerably for a given foraging effort. This was demonstrated on two observation days: on 4 August 1993, colony I collected 0.63 kg dry weight made up of 165,000 loads consisting exclusively of *Dipteryx* flower parts. In comparison, the total accumulated in the nest on 2 February 1994 was 1 kg of biomass consisting of almost the same number (155,175) of *Ficus* stipules and leaf fragment loads.

Foraging activity counts could be converted to biomass intake and used for models estimating the total daily vegetation harvest (see Sect. 8.4.1) if a site was independent and weight of harvested items existed. For *A. colombica* both the average fragment weight cut by colony I during 1993/1994 (5.51±2.5 SD

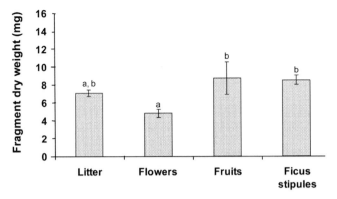

Fig. 31. Dry weight of nongreen materials collected by ants of colony I (litter: $n=44$, flowers: $n=29$, fruits: $n=22$, *Ficus* stipules: $n=28$). Dry weight of resource classes was different (ANOVA: df=3, 118; $p=0.005$; F=4.496) with significant differences only between dry weight of flower parts and both *Ficus* stipules and fruits (Tukey's multiple comparison; *different letters* above the columns indicate significant differences at $p=0.001$). *Litter* refers to leaf litter, twigs, bark and other miscellaneous items collected. (Total number of samples analyzed: 45, consisting of a total number of 6896 fragments.)

mg) as well as the mean value described by Lugo et al. (1973) in Costa Rica (7.99 mg) were within the range of the 49 colonies sampled during 1997/1998 (6.51±1.91 SD mg). On the other hand, the differences in mean fragment size seem to be greater if different *Atta* species are compared. For example, as an average weight of several *Atta* species, Fowler et al. (1990) found 9.2 mg per fragment, and Rockwood (1973) stated that *A. cephalotes* generally cuts larger fragments than *A. colombica*. Thus, species-specific differences may explain most of the discrepancies in mean load sizes among *Atta* colonies. Differences between colonies of the same species may also be due to differences in mean worker size (c.f. Wetterer 1990), and differences in vegetation composition between colony sites (Fowler et al. 1990). In conclusion, the conversion of leaf-cutting activity into biomass data by means of a "generalized leaf-cutting ant load weight" is a legitimate approximation especially if applied with species-specific fragment weights.

8.3 Diurnal Course of Foraging Activity

8.3.1 The Typical Pattern and Its Variability

Figure 32A shows the average leaf harvesting activity curve for colony I based on the results of 44 counting days during 1 year. The data confirm the pattern already known for *A. colombica* in Panama (Lutz 1929; Hodgson 1955; Weber

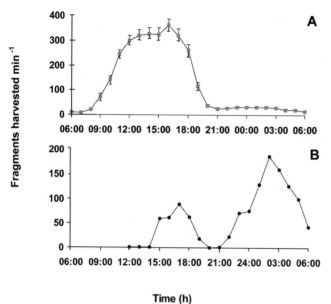

Fig. 32. Average daily cycle of foraging activity for colony I derived from $n=44$ survey days (**A**); *error bars* represent ±SE. (**B**) Atypical foraging activity for the same colony on 27 September 1993

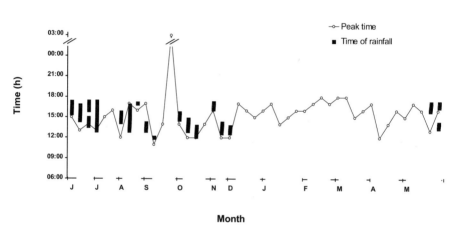

Fig. 33. Peak of daily foraging activity compared to time of rainfall on survey days (only rainfall between 12:00 and 17:00 h considered, i.e., events within the period of maximum harvest rates). *Ticks* on the *abscissa* represent the beginnings of the labeled months (four values per month, except for November). (Wirth et al. 1997)

1972b). Foraging activity outside the nest started at dawn, remained high during daylight hours, and practically ceased after dark. However, counts at 24 other *A. colombica* colonies at different randomly selected times of the year on BCI revealed some exceptions to this rule (Table 13). Almost half of the observed colonies showed considerable night-time activities, four of them with nocturnal foraging rates >50% of the daily maximum. One colony seemed to be exclusively night-active during the time of observation with a

Table 13. Diel patterns of foraging activity of 24 randomly selected colonies of *A. colombica*

Foraging activity period	No. of Colonies
Daytime	13
Primarily daytime[a]	6
Day and night[b]	4
Nighttime	1

[a] Nocturnal activities min^{-1} (measured between 23:00 and 3:00 h) 25–50 % of the maximum daylight activity.

[b] Nocturnal activities min^{-1} (measured between 23:00 and 3:00 h) >50 % of the maximum daylight activity.

peak activity around midnight. Typically, the diurnal activity cycle of colony I was characterized by fairly constant patterns throughout the year, with low levels at night and peaks between 2:00 and 5:00 p.m. (Fig. 33). Nevertheless, occasional deviations from the normal pattern can occur, as demonstrated on 27 September 1993, when foraging was severely limited during daylight hours as a consequence of hostile interactions between arboreal carpenter ants (*Camponotus sericeiventris*) and *Atta* foragers (Fig. 32B). During daylight hours, *Camponotus* workers guarded a large host tree (*Hura crepitans*), preventing the majority of the colony's foraging force from ascending and harvesting until the evening. However, at night, in the absence of *Camponotus*, an exceptional nocturnal harvest of that tree took place, resulting in a final input of *Hura* leaves of about 70 % of the days harvest. Interspecific hostilities between *Atta* and *Camponotus* ants frequently took place on several host trees of many colonies on BCI (e.g.,>10 dominant trees of colony I) usually being accompanied by limited leaf-cutting ant access to the respective tree (pers. observ.).

The wide variety of pronounced activity patterns found between colonies is a frequently discussed aspect of leaf-cutting ant ecology. Variations in the circadian activity patterns of many species and in several regions have primarily been attributed to climatic factors (e.g., Waller 1986; Fowler and Robinson 1979). Foraging was observed to be either predominantly diurnal or nocturnal (Lutz 1929; Hodgson 1955; Cherrett 1968; Lewis et al. 1974a; Wetterer 1988) and has even been described to differ between neighboring colonies (Lewis et al. 1974b; Pilati and Quiran 1996). Although a strong trend for diurnal activity was found for most *A. colombica* colonies on BCI, the results of the present study confirm this wide range of variation. None of the many attempts to explain colony-specific heterogeneity of activity patterns has been particularly successful. Neither those seeking to link foraging rhythm to physical or climatic factors (Hodgson 1955; Cherrett 1968; Lewis et al. 1974a; Pilati and Quiran

1996), nor those trying to relate it to biotic factors produced final evidence. For example, parasitic phorid flies (Diptera: *Phoridae*) were shown to influence *Atta* foraging patterns (Feener and Moss 1990; Orr 1992; Braganca et al. 1998), but since phorid flies are not likely to restrict their parasitic activity to the territory of single *Atta* colonies, this cannot adequately account for asynchronous periodicity of adjacent colonies. As described above, carpenter ants (*Camponotus sericeiventris*), excluding *Atta* foragers from their defended host tree, were identified as a new reason for variations in foraging activity patterns of *Atta* colonies. Hence, differences in the abundance of defended host plants between colonies could theoretically account for foraging cycle differences among different colonies. However, since the foraging pattern of the respective colony was altered only for a rather short period of time, there is no clear evidence that such phenomena play a major role in this connection. Future research on colony-specific harvesting patterns should, therefore, focus on the following aspects: (1) since several arboreal ant species have been identified to protect their host plants against leaf-cutting ants (e.g., Wetterer 1994a; Vasconcelos and Casimiro 1997), foraging patterns should be further related to these interactions; (2) the potential of the cumulative deterring effects for foraging as triggers, particularly for the early determination of a colony's foraging activity cycle should be assessed; (3) internal physiological rhythms of the colonies such as nutritional requirements in combination with heterogeneous resource availability should be considered as a potential cause for changes in foraging patterns (cf. Farji-Brener 1993) and (4) territorial shapes of adjacent colonies should be studied with regard to their foraging activity patterns, since partitioning of overlapping territories may be enabled through foraging in different time niches. (During the present study, a nocturnally foraging *Acromyrmex octospinosus* colony was foraging wholly within the area of the diurnally active colony I.)

8.3.2 The Role of Precipitation

Since ant activity outside the nest typically ceases during heavy rain events (e.g., Hodgson 1955; pers. obs.), the efficiency of leaf-cutting ant foraging should be strongly affected by length and intensity of rain events. The amount of precipitation during the daily activity periods of colony I accounted for ca. 50 % of the variability of total biomass intake during days with rainfall events (Fig. 34). As illustrated by Fig. 33, variation in time and duration of rainfall events probably affected the strength of this relationship. If extended periods of heavy rain occur during or immediately before the foraging peak, the daily total of collected plant material may be substantially reduced. However, if heavy rain occurs several hours before or well after the normal peak foraging time, subsequent re-activation of foraging trails is possible, and daily forage totals would be only moderately reduced (Wirth et al. 1997).

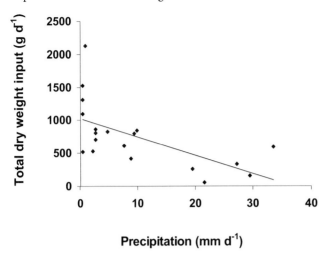

Fig. 34. Relationship between total dry weight of plant material collected by colony I and total precipitation during daylight hours ($r=-0.614$, $p<0.05$, $n=19$; only days with rainfall considered; climatic data provided by the Terrestrial Environmental Science Program, TESP, STRI, Panama)

8.4 Rapid Estimation of Harvesting Rates

8.4.1 Estimating Colony Harvest

Analysis of the daily harvest of leaf fragments revealed a highly significant relationship between the leaf harvesting rate during the afternoon peak of activity and the total daily leaf harvest for colony I (Fig. 35A: $r^2=0.78$; $p<0.001$; $n=45$; peak values excluded). The strength of the correlation reached its maximum at 4:00 p.m., but was still significant at other afternoon hours. An analogous analysis of the 15 data pairs from the activity counts of colony II yielded a similarly strong correlation ($r^2=0.79$; $p<0.001$; $n=15$). Further, regression relationships established separately for the 17 most frequently harvested plant species of colony I ($r^2=0.88$; $p<0.001$; $n=21$) indicated that daily harvests of leaves from particular plant species also follow the same rule. As can be seen in Fig. 35B, the pooled harvest data from colonies I and II fall on the same regression line ($r^2=0.87$, $p<0.001$, $n=81$) indicating that this correlation may be colony independent. To test the hypothesis that this regression is of general validity, data from published studies of Hodgson (1955) and Lugo et al. (1973) and additional activity counts from colonies III–V on BCI and a nocturnal *Atta cephalotes* colony located in the Parque Metropolitana near Panama City were analyzed. Figure 36 shows that the regression estimates from these colonies, in fact, correspond well with the observed daily intake rates ($r^2=0.929$; $p<0.001$, $n=9$). The predicted values appear to be less accurate for colonies with very low daily harvests (e.g., BCI colonies III and V). The high correspondence (deviation=0.87 and 5.75 %) found for the nocturnal foraging *Atta cephalotes* colony (peak foraging time 21:00–22:00 h.) further

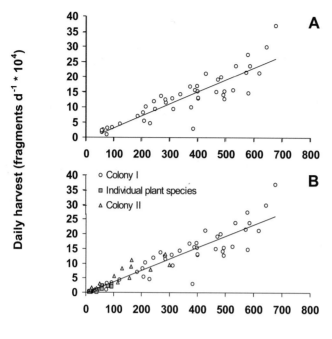

Fig. 35. Relationship of total number of green leaf fragments collected per day and the number of leaf fragments harvested per minute at the peak of daily activity. In **A**, data are derived from activity measurements of total green material of colony I. In **B**, the data sets of colony II and two daily courses of 17 plant species counted at colony I are combined. (Regression equations in **A** y=0.0399x −0.7719; in **B** y=0.039x−0.2128.) (Wirth et al. 1997)

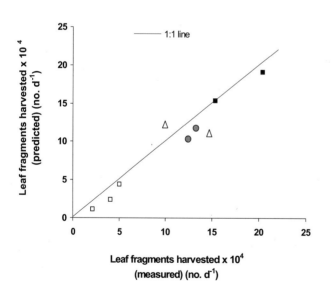

Fig. 36. Correspondence between measured (estimated from hourly counts) and predicted (using the linear regression model from Fig. 35B) total daily harvest rates (total number of collected green leaf fragments per day). Data from *A. colombica* colonies III, IV, and V on BCI (*open squares* 1 day each), one colony of *A. cephalotes* in Parque Metropolitana, Panama City, Panama (*filled squares* 2 days), and published studies of *A. colombica* by Hodgson (1955) and Lugo et al. (1973) (*filled circles* and *open triangles* 2 days each). (Wirth et al. 1997)

indicates that this relationship may hold for other *Atta* species as well. A similar regression model derived from the correlation between the daily maximum intake rate and the total daily intake of all harvested plant parts for colony I (i.e., green plus nongreen fragments), was only slightly less powerful ($y=396.52x+9713.3$, $r^2=0.5574$; $p<0.001$, $n=45$). Using daily activity counts from 23 *Atta colombica* colonies on BCI, this model led on average to a 15% ($\pm23\%$ SD) underestimation of the measured daily intake (Fig. 37).

8.4.2 Estimating the Annual Harvest of the *Atta Colombica* Population

Atta colombica is characterized by its superficial colony refuse-dump (Haines 1978; Plate 10). While in most other species of leaf-cutting ants the refuse is buried in subterranean chambers around and below the fungus chambers (Stahel and Geijskes 1939; Moser 1963; Jonkman 1980b; Hölldobler and Wilson 1990), *A. colombica* assigns a single trail exclusively to ants carrying particles of refuse material out of the nest to the dump site where they typically drop the refuse from above after climbing on the trunk of a tree, a liana, or a rock (Plate 11). While the number of ants carrying plant fragments into the nest varies throughout the day (see Sect. 8.3.1), data from Hodgson (1955) and also our own preliminary observations suggested a rather constant diurnal refuse dump rate.[*] To test whether this behavior can be generalized for *A. colombica* and to investigate whether there is a correlation between the refuse

---

[*] Hodgson (1955) described the observed leaf-cutting ant species on BCI as *A. cephalotes*. Meanwhile, it is well known that *A. cephalotes* dumps its colony refuse exclusively belowground. Thus, Hodgeson obviously misidentified *A. colombica* as *A. cephalotes*.

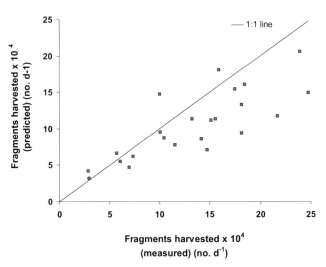

Fig. 37. Correspondence between measured (estimated from hourly counts) and predicted total daily number of all biomass fragments collected per day (linear regression model described in the text). Data from *A. colombica* colonies III to XXV on BCI (1 day each). The regression equation is $y=0.5953x+25712$ ($r^2=0.638$, $p=<0.001$, $n=23$)

dump rate and the number of active foragers, activity counts of laden ants on foraging trails as well as on refuse trails of colonies of various sizes were carried out for 24-h intervals. Five-minute counts were taken at regular intervals of 1–1.5 h to determine the number of laden ants entering a nest with plant fragments or leaving with refuse particles per day. From these counts the total daily number of foragers and refuse carriers was extrapolated. Diurnal counts were performed on 20 nests of various sizes throughout the years 1996 and 1997. To determine the total refuse-dumping activity of the *A. colombica* population, all existing colonies were visited monthly from July 1997 until July 1998 and population counts were conducted as described in Section 8.1.1, but with a counting interval of 5 min.

As predicted by Hodgson (1955), the diurnal refuse-dumping rate was relatively constant in all examined colonies, whereas the number of ants carrying plant fragments into the nest varied by the time of day (Fig. 38). The mean coefficient of variation of refuse-dumping activity counts was determined as 9.3±6.1 % (mean ± SD, n=22 diurnal courses), while the coefficient of variation of foraging-activities was found to be more than ten times higher (98.3±30.9 %; mean ± SD, n=22 diurnal courses). The total daily number of harvested plant fragments carried into the nest strongly correlated with the number of refuse particles removed from the nest during 24 h (r^2=0.767, p<0.0001, n=22; Fig. 39). Thus, it should be possible to estimate the number of harvested fragments per day directly from short time counts of refuse dump activity. This hypothesis was tested by comparing the measured foraging activity from 22 available diurnal counts with the estimated activity from the respective refuse counts taken randomly throughout the same days and cal-

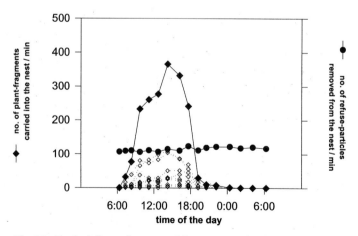

Fig. 38. Typical diurnal course of fragment and refuse loads of an entire medium-sized *A. colombica* colony. The individual courses of the nine major trails of this colony are indicated by *open symbols*

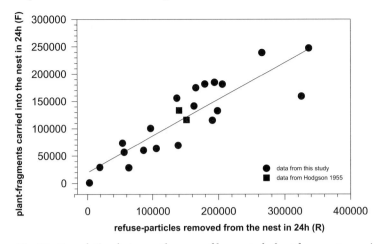

Fig. 39. Correlation between the sum of harvested plant fragments carried into the nest and the sum of refuse particles removed from the nest over 24 h in 22 *A. colombica* colonies on BCI and from a literature reference. The regression equation is $y=0.6697x+20472$ ($r^2=0.767$; $p<0.001$, $n=22$)

Table 14. Harvest rates of *A. colombica* estimated in 100 ha of late successional forest on BCI

Parameter	Estimate	Unit
Number of foragers	1.55×10^9	(ants year^{-1})
Annual leaf area harvested	12.2	(ha year^{-1})
Annual leaf area harvested per ground area	1217	(m^2 ha^{-1} year^{-1})
Daily leaf area harvested per ground area	3.33	(cm^2 m^{-2} day^{-1})
Leaf area harvested per colony	2558	(m^2 colony^{-1} year^{-1})
Total biomass harvested	11576	(kg year^{-1})
Annual biomass harvested per ground area	115.7	(kg ha^{-1} year^{-1})
Daily biomass harvested per ground area	0.0317	(g m^{-2} day^{-1})
Biomass per colony	243.2	(kg colony^{-1} year^{-1})

culated according the correlation given in Fig. 39. In all ten cases, the estimated foraging activity closely reflected the measured diurnal foraging activity from these 22 diurnal counts (range of r^2: 0.674–0.791). After establishing and verifying this correlation and based on the data shown in Table 12 and on additional literature information on the leaf area index of the forest (Leigh and Smythe 1978; Wirth 1996), it was now possible to calculate the annual harvest rate within the focal area. The results of these calculations are given in Table 14.

8.4.3 General Suitability of Linear Regression Models for Estimating Harvest Rates

The proposed use of regression equations is an approach to quantify the harvests of leaf-cutting ants using instantaneous rates of (1) foraging or (2) refuse-dumping activity. Whereas foraging rates have to be measured at the peak of daily activity, refuse dumping is a rather constant diurnal process allowing counts at any hour during the day. Measuring biomass intake by means of short-term activity counts of carried fragments may provide a time- and labor-efficient new tool for researchers to make good predictions of daily biomass input into leaf-cutting ant nests, either on the level of total harvest as well as for different resource types such as leaves, nongreen items or even individual plant species (cf. Wirth et al 1997). Merely the determination and verification of the colony-specific activity peak would be needed prior to the application of the method and could be accomplished with a few days of observation. Inaccuracies are only to be expected if normal activity patterns are persistently disturbed by rainfall. In contrast, the use of refuse-dumping counts may be particularly useful for ecosystem level studies as they allow the determination of foraging activities of many colonies across large areas and over a long time to be quantified. Whereas the foraging activity method is suitable for all leaf-cutting ant species without reservation, the refuse method, as it depends on external refuse dumping, can only be applied to *Atta colombica* and very few other species.

9 The Trail System

Foraging trails make it easier for workers of social insects to find resources after leaving the nest (Hölldobler 1977; Shepherd 1982; Fowler and Stiles 1980). The way in which the food sites are used is thus a consequence of the distribution of the food in space (Carroll and Janzen 1973). Apart from this partitioning effect (i.e., the location of the harvesting sites within the colony territory), trails have been attributed to an aggression-reducing function between neighboring colonies whose resource-containing areas overlap (Vilela and Howse 1986; Farji Brener and Sierra 1993). According to Hölldobler and Lumsden (1980), the gradually shifting trunk trails and the area immediately around the nest are regarded as a "core area" which is the actual territory to be defended. Thus, the trails would obviously reduce the chance of aggressive confrontation between adjacent colonies. However, this hypothesis has not yet been conclusively proven for leaf-cutting ants.

There is a general consensus on the central role of trunk trail systems (Plates 4, 5) for the foraging of leaf-cutting ants. The use of physical foraging trails by leaf-cutters and many other ants is an important mechanism that serves to facilitate the encounter of resource items (Shepherd 1982; Hölldobler and Wilson 1990) and can be considered as a "physical memory" of resource patches previously encountered. They enhance foraging efficiency by increasing foraging speed four- to tenfold compared to uncleared ground (Rockwood and Hubbell 1987) and they are postulated to encompass foraging territories that serve to protect the colony's resources from competitors (Fowler and Stiles 1980; Hölldobler and Lumsden 1980).

In contrast to these benefits, trail construction and maintenance may add a significant proportion to the overall energy investments for resource acquisition (Lugo et al. 1973; Shepherd 1982). However, in a recent study on the costs of trail construction and maintenance in *Atta colombica* on BCI, Howard (2001) found that energy requirements are small relative to the harvest rate and worker force of a leaf-cutting ant colony, suggesting that costs do not significantly constrain trail construction. Further, it has been suggested that foraging is trail-centered, the search effort for high quality resources being restricted to a close vicinity of the trunk trails (Fowler and Robinson 1979;

Shepherd 1982; Farji-Brener and Sierra 1988; Vasconcelos 1990b). Despite the fact that trunk trail systems imply a great deal of theoretical importance for the understanding of foraging in leaf-cutting ants, the available field data are limited and not completely conclusive to confirm current foraging models (Fowler and Robinson 1979; Lighton et al. 1987; Rockwood and Hubbell 1987). Up to now, little attention has been given to the origin, maintenance and permanence of the trails or their variability in time and space (Shepherd 1985; Farji Brener and Sierra 1993). Therefore, a detailed monitoring of the activities of ant foraging columns of a *Atta* colony was conducted for a whole year during the present study.

9.1 Mapping and Monitoring of Foraging Trails

Sources of all leaves carried into colony I and the corresponding foraging trails were monitored at 2–3 day intervals from June 14, 1993 to May 29, 1994, which resulted in a total of 126 observation days. On each monitoring day all current and new foraging trails were flagged at intervals of 7.5 m (measured along the ant trail) and followed up to the cutting site or the location where ants ascended into the canopy (hereafter both are referred to as 'foraging sites'). The flags were labeled with notations on the distance to the beginning of the trail and the degree of trail ramification. The notation was based on a hierarchic arrangement of the trunk trails which ramified into first, second, third, etc. order trails. Foraging sites were separately labeled with a combination of the trail notation and chronologically assigned letters in the order of their first record. Both trails and foraging locations were recorded for later mapping by noting down the compass bearing and distance from each newly identified cutting site to the last defined branching point of the trail.

9.2 Annual Course of Foraging Trail Establishment

Maps showing the total active trails and corresponding foraging sites of each month of the observation period revealed major variability in both number of resources currently exploited and the degree of dispersion of foraging (Figs. 40, 41). Nevertheless, it is obvious that the foraging activities concentrated along four main foraging axes extending from the central nest roughly towards west, northeast, southeast, and south. These main routes were retained throughout the entire observation period, being resumed even after a long-lasting interruption of the foraging activities [e.g., trails II (south) and III (west) during February or trail IV (northeast) during April and May 1994; Figs. 40, 42]. A cumulative presentation of the trail monitoring data,

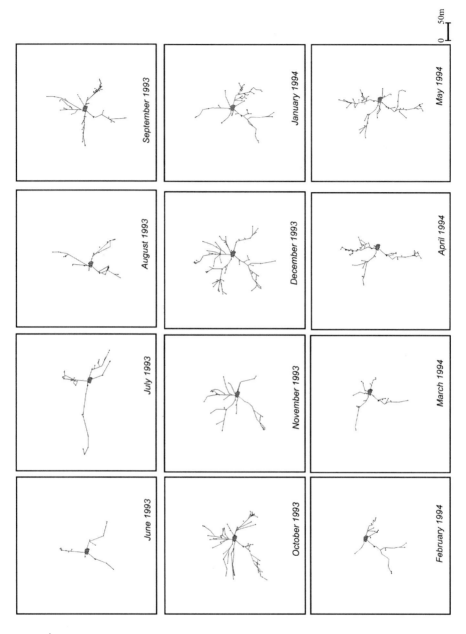

Fig. 40. All foraging sites (*dots*) and trails (*lines*) of colony I separately mapped for each month during a complete year of observation (June 1993–June 1994). The nest is indicated by the central *stippled polygon*

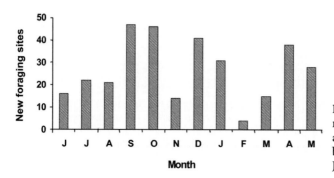

Fig. 41. Number of newly established foraging locations between June 1993 and June 1994

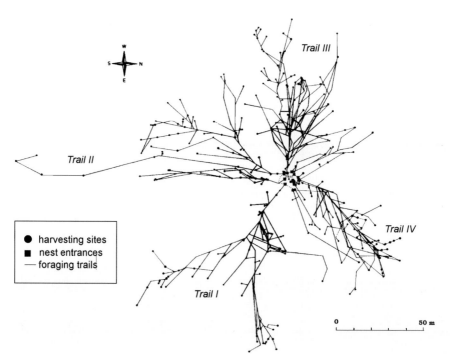

Fig. 42. All foraging sites and trails used by colony I between June 1993 and June 1994. Cumulative presentation of the information shown in Fig. 40

showing all foraging trails and sites ever established in the course of the year is shown in Fig. 42. Maximum foraging distances among the four trunk trails averaged about 80 m, the furthest leaf source being located about 158 m south of the nest. Strikingly, no foraging trails were found between the four principal axes.

9.3 Frequency of Trail Establishment Throughout the Year

Many of the leaf sources were accessed by third and fourth order trails, like small single trails to additional entrance holes (i.e., not corresponding to persistent trunk trails). These trails were temporary and often used for a single day only. On average, 27 new ascending locations were established per month, i.e., almost one per day. However, this does not necessarily correspond with the number of new source plants, since a single canopy access trail can be used to simultaneously collect material from several plant individuals (see below). The number of foraging sites varied considerably over time, reaching distinct peaks in September and October, when up to 47 new ascending sites were found per month, whereas exceptionally few (4) were found during the dry month of February (Fig. 41). The total number of active trails was generally slightly higher than the number of newly established trails for a particular month and positively correlated with the rates of mean daily harvest measured at the same colony (Fig. 43; see also Fig. 25). Regarding the high significance of this linear relationship (r^2=0.63, p=0.02, n=12), the number of present trails possibly serves as a general indicator for the efficiency of leaf-cutting ant harvesting.

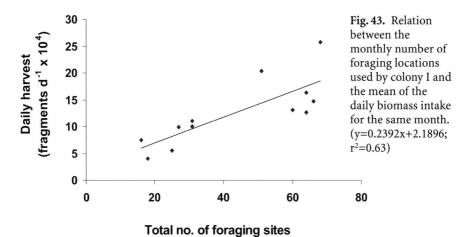

Fig. 43. Relation between the monthly number of foraging locations used by colony I and the mean of the daily biomass intake for the same month. (y=0.2392x+2.1896; r^2=0.63)

9.4 Spatial and Temporal Patterns of Foraging

Monitoring of foraging trails and cutting sites over a period of 1 year yielded some new insights in the nature of branch trail formation. It could be shown that the long-term collective foraging effort of the colony was allocated to four main foraging areas located at roughly the same distance apart from each other. Thus, the potential colony territory was divided into almost equal sectors, leaving out large areas between the trunk trails. This implies that large sections of the potentially available forest area were not used by the ants in the course of 1 year. A subsequent survey of host plant use during the following year (Weigelt 1996) revealed that the layout of the foraging routes had still not noticeably changed, i.e., the previously "ant-free" zones were still not being exploited. In addition, 59 of the 64 foraging sites registered during this time (April and May 1995) were located within the foraging area of 1993/1994, and 48 of these were identified as having been used as leaf sources during the 1993/1994 study. The present results as well as observations of Vasconcelos (1990b) on *A. cephalotes* and *A. sexdens* contradict the scarce published information. Farji-Brener and Sierra (1993) reported a rotational pattern of foraging activity for *Atta colombica* on BCI for 7 days of observation and interpreted this as a general and unpredictable harvest, covering the whole area surrounding the nest. This discrepancy may be explained by the fact that the studies of Farji-Brener and Sierra were performed in October, a period of increased trail modification activity, as shown by the annual course of trail construction (see Fig. 41).

The high significance of the correlation between daily biomass intake and the number of foraging trails (Fig. 43) underpins the suggested link between the quality of the available resources and the degree of harvest specialization (Blanton and Ewel 1985; Shepherd 1985; Rockwood and Hubbell 1987; Vasconcelos and Fowler 1990). Foraging patches were changed much less frequently during the harvesting of nutrient-rich flower parts, fruits and large *Ficus* stipules in February and March (see Figs. 40, 41). If it is assumed that the absence of trail establishing activity implies lower harvest diversity, it can be concluded that higher substrate quality leads both to greater specialization of the workers on specific food sources and to longer utilization of a specific foraging patch. This is in line with the elementary predictions made by theories of optimal foraging (see Pyke et al. 1977). The lack of young, high quality leaves at the climax of the rainy season (Leigh and Windsor 1982; Hubbell et al. 1984) and the corresponding high frequency of switching to new resources in September and October support this view.

9.5 Foraging Activity in the Canopy

The maps shown in Figs. 40 and 42 provide only limited information about the actual location of the leaf sources in the forest canopy. Foraging ants often ascend by any available trunk or liana and then proceed to the actual leaf source once they are in the canopy. Following a selection of such "canopy trails", it became obvious that travel distances around 20 m are not unusual within the topmost stories of the forest. Thus, leaves of several species may be carried down a single route. In most cases, however, the leaf source corresponded with the ascending location.

9.6 Estimating Foraging Area

The trail monitoring information could also be used for a rather realistic estimation of the actual size of the foraging area. Total areas enclosed by the contours in Fig. 44 were determined as 2712 m² for trail I (SE); 2597 m² for trail II

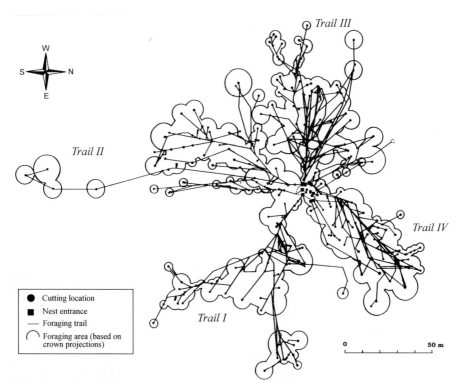

Fig. 44. Outline of foraging area of colony I derived from regression estimates of crown projection areas (see also Fig. 50)

(S), 2640 m² for trail III (W) and 2409 m² for trail IV (NE), thus resulting in a total foraging area of 1.03 ha for colony I.

The size of the foraging territory of *Atta* colonies is an important reference variable for all studies on the impact of the ants on the vegetation. At a given consumption rate, their direct effects must decrease as the size of the area increases. Most regrettably, an assessment of this variable is missing in many publications on biomass harvest or other ecosystem impacts of leaf-cutting ants (e.g., Hodgson 1955; Blanton and Ewel 1985). In the few publications where the size of the foraging area was in fact estimated, it was done by simply joining the ends of existing trail systems and using elementary geometrical shapes (rectangles, ellipses, etc.) for the calculations (Lugo et al. 1973; Haines 1978; Rockwood and Hubbell 1987; but see Vasconcelos 1990b). Fig-

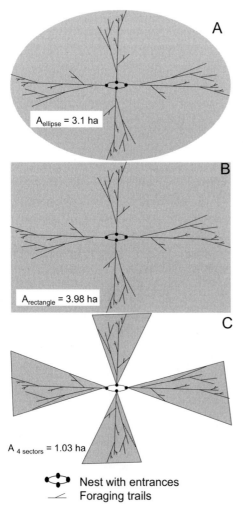

Fig. 45A–C. Three approaches to estimate the size of leaf-cutting ant foraging territories on the basis of long-term mapped trunk trail systems. Connecting the terminal points of existing trails, the total area can be calculated by the assumption of elliptical (**A**), or rectangular (**B**) shapes. In **C**, the foraging sectors actually visited by workers during a given time period are separately outlined and summed (see present study). The given area values (*A*) correspond to the *gray shapes*

$A_{ellipse}$ = 3.1 ha

$A_{rectangle}$ = 3.98 ha

$A_{4\ sectors}$ = 1.03 ha

Nest with entrances
Foraging trails

ure 45 compares the results of two of these methods with the estimation method of the present study. It is evident that the area obtained by considering the real situation (i.e., separate foraging sectors, Fig. 45C) results in values that are only one quarter to one third of the areas based on rectangular or elliptic shapes. Therefore, for future more realistic impact studies, the use of the approach shown in Fig. 45C is strongly recommended.

9.7 Trunk Trail Systems as Determinants of Foraging Patterns

As a synthesis of published observations and trail organization models (e.g., Fowler and Stiles 1980; Hölldobler and Lumsden 1980; Shepherd 1982; Reed and Cherrett 1990) and the results presented here, the following hypothesis for the establishment of trails and the foraging strategy of *Atta* species can be proposed: the initial foraging experience acquired by a young colony determines its spatial foraging pattern in the future. In other words, the orientation and dispersion of the trail system of a given colony are primarily consequences of prior spatial allocation of foraging effort and less of the overall distribution of resources around the nest. Once established, the rough locations of the foraging trails do not change in the long-term of a colony's lifetime. However, within such an area-restricted foraging strategy, leaf-cutting ants are able to exhibit a spatially and temporally highly variable response due to the patchy distribution of high quality resources in time and space. The above hypothesis is supported by the following results: (1) a considerable spatial persistency of primary foraging sectors and "ant-free" zones after 1 year of observation; (2) confirmation of the general pattern of the trail system and repeated harvesting of the same plant individuals (at the same time of the year) in the subsequent year; (3) according to earlier observations (Cherrett 1968), workers do not leave previously used trunk trails despite the fact that highly palatable plant specimens would be available close to the nest and the trunk trails [for instance: foraging in the area surrounding trail III was limited to about a third of the tree individuals potentially accessible, although the colony generally preferred some of the avoided trees species (e.g., *Poulsenia armata*, *Quararibea astrolepis*, *Macrocnemum glabrescens* and *Dipteryx panamensis*)]; (4) food transport along cleared trails is faster (and therefore cheaper) than over the littered forest floor, and the energetic cost of establishing a new trail is higher than that of maintaining an existing one (Shepherd 1982; Lighton et al. 1987; Howard 2001); and (5) observations and studies with pit-fall traps (Fowler and Robinson 1979; Shepherd 1982) showed that foraging effort and food transport are concentrated on and near existing trails; new trails are laid down mainly from the ends of trails in current use and are limited by a maximum "permissible" distance from the nest (Cherrett 1983; Reed and Cherrett 1990). This hypothesis does not contradict a strategy proposed by several authors in which both foraging trails and territories are a consequence of the heterogeneity of patchy distributed resource qualities (e.g., Fowler and Stiles 1980; Rockwood and Hubbell 1987). The variability of trail persistence and the diversity of the food spectrum revealed that such a resource-controlled foraging behavior within the sectors is quite conceivable.

In contrast, the equal distribution of foraging activities over the whole of the potential area to avoid overexploitation of single resource plants as required for the "resource conservation" theory (Cherrett 1968; Rockwood 1975), like the results of Farji-Brener and Sierra (1993), are incompatible with this model. A recent study on *Atta sexdens* in French Guyana revealed that young colonies in fact showed lower degrees of mean trail persistence and trail branching as well as increased homogeneity of spatial exploration of their home range (Wirth et al. 2001a).

10 Host Plant Selection

Generally, host plant selection of leaf-cutting ants and particularly the mechanisms of selective foraging in *Atta* ants have been extensively studied with laboratory colonies as well as with natural colonies in the field using pick-up assays as method for both (e.g., Cherrett 1972a,b; Hubbell and Wiemer 1983; Howard 1987, 1988; Vasconcelos and Cherrett 1996; Salatino et al. 1998). However, long-term descriptive field studies are rather scarce (Rockwood 1975, 1976; Shepherd 1985; Wirth et al. 1997). Thus, for a better understanding of the factors influencing food choice under natural conditions throughout the year, it was necessary to investigate host plant selection and to determine the degree of polyphagy of *A. colombica* throughout an entire year.

10.1 Measuring Electivity

The resources being harvested by colony I throughout the observation year (for details on species identification, see Wirth et al. 1997) were identified simultaneously to the activity monitoring (see Chap. 9). Since the plants used for ascending and descending into and from the canopy were not necessarily identical to the source plants of the fragments harvested (see Chap. 9.5), the exact origin of the fragments always had to be determined carefully. For this, the ant trails were either directly tracked in the canopy with binoculars or the harvested fragments were compared with leaves of the plants in the surroundings of the ascending location. In order to gain information on the harvest preferences of the ants from colony I, the following relationships between the harvested plants and the total number of potentially available plants were established. For this, data of the vegetation inventory of the study plot (see Chap. 4) were used together with harvest data of trunk trail III which was largely situated within the study plot.

First, the inverse of Simpson's index D (Simpson 1949) was used to express the diversity of items in the ants harvest:

$$D = \frac{1}{\sum_{i=s}^{s}(p_i)^2} \tag{5}$$

where S is the number of species, and p_i is the proportion of species i in the community. Simpson's index is particularly suitable in this context since its inversion is a measure of "equally used" resources representing the same level of diversity. The corresponding evenness E, was used as a measure of homogeneity of species abundance. It was expressed by the relationship between the substrate diversity and the maximum diversity for the given number of species:

$$E = \frac{D}{1 - \frac{1}{S}} \tag{6}$$

Second, harvest preferences comparing usage and availability of food types were measured by means of the electivity index (Ivlev 1961):

$$El_i = \frac{r_i - n_i}{r_i + n_i} \tag{7}$$

where El_i is the electivity index for species i, r_i is the percentage of species i in the harvest, and n_i is the percentage of species i in the environment.

10.2 Harvest Preferences of Colony I Throughout the Year

Table 15 lists all identified plant sources together with information on plant organs harvested, growth form, and plant functional type (regeneration strategy). Throughout the observation year, foraging workers from colony I harvested material from 126 plant species (15 of which could not be identified due to difficulties in obtaining sufficient leaf samples).

10.2.1 Characterization of the Harvested Species

Growth form: 64.1% of 117 forage species considered (i.e., where growth form was determinable) were trees and shrubs and about one third (29.9%) were lianas. Epiphytes and hemiepiphytes (3.4%) as well as herbs (1.7%) were of little importance. Except for *Philodendron*, the epiphytic fern *Elaphoglossum sporadolepis*, the understory palm *Chamaedorea wendlandiana* and a herb-like *Calathea* species, all harvested species were dicotyledons.

Table 15. Plant species and resource types collected by colony I during 1 year of observation. Plants are characterized by classifications of growth form (cf. Croat 1978) and regeneration strategies (according to Condit et al. 1996a; R. Condit, pers. comm.; S.P. Hubbell, pers. comm.; P.D. Coley, pers.comm.; Dalling et al. 1998)

| No. | Species | Family | Resource type | | | | Growth[a] form | Regeneration strategy[b] |
			Leaf	Flower	Fruit	Other		
1	*Acalypha diversifolia*	Euphorbiaceae	x				s	i
2	*Acalypha macrostachya*	Euphorbiaceae	x				s	p
3	*Alchornea costaricensis*	Euphorbiaceae	x				t	p
4	*Allophyllus psilospermus*	Sapindaceae	x				t	s
5	*Alseis blackiana*	Rubiaceae	x				t	s
6	*Anacardium excelsum*	Anacardiaceae	x	x	x		t	s
7	*Anemopaegma chrysoleucum*	Bignoniaceae	x				l	
8	*Annona spraguei*	Annonaceae	x				t	p
9	*Apeiba membranacea*	Tiliacaeae	x				t	p
10	*Apeiba tibourbou*	Tiliaceae	x				t	p
11	*Arrabidaea florida*	Bignoniaceae	x				l	
12	*Bombacopsis sessilis*	Bombaceae	x				t	p
13	*Brosimum alicastrum*	Moraceae	x	x	x		t	s
14	*Byrsonima crassifolia*	Malphigiaceae	x	x			t	p
15	*Byttneria aculeata*	Sterculiaceae	x				s/l	
16	*Calathea* sp.	Marantaceae	x				h (m)	
17	*Calophyllum longifolium*	Guttiferae (=Clusiaceae)	x				t	
18	*Capparis frondosa*	Capparidaceae	x				s	s
19	*Casearia aculeata*	Flacourtiaceae	x				t	s
20	*Casearia arborea*	Flacourtiaceae	x	x			t	p
21	*Cassipourea elliptica*	Rhizophoraceae	x				s/t	s
22	*Cavanillesia platanifolia*	Bombacaceae	x				t	s
23	*Cecropia insignis*	Moraceae	x				t	p
24	*Cecropia obtusifolia*	Moraceae	x		x	Stipules	t	p
25	*Ceiba pentandra*	Bombacaceae	x				t	p
26	*Chamaedorea wendlandiana*	Palmae	x				t (m)	s
27	*Clusia uvitana*	Clusiaceae	x				he	
28	*Coccoloba parimensis*	Polygonaceae	x				l	
29	*Coccoloba* sp.	Polygonaceae	x				l	
30	*Combretum decandrum*	Combretaceae	x				l	
31	*Cordia alliadora*	Boraginaceae	x				t	p
32	*Didymopanax morototoni*	Araliaceae	x				t	p
33	*Dipteryx panamensis*	Leg.-Pap.	x	x	x		t	i
34	*Doliocarpus* sp.	Dilleniaceae	x				l	
35	*Doliocarpus olivaceus*	Dilleniaceae	x				l	
36	*Doliocarpus dentatus*	Dilleniaceae	x				l	
37	*Doliocarpus major*	Dilleniaceae	x				l	
38	*Doliocarpus multiflorum*	Dilleniaceae	x				l	
39	*Elaphoglossum sporadolepis*	Polypodiaceae	x				e	
40	*Entada monostachya*	Leg.-Mimos.	x				l	
41	*Erythroxylum panamense*	Erythroxylaceae	x				t-s	s
42	*Eugenia oerstedeana*	Myrtaceae	x				t	s

Table 15. (*Continued*I

No.	Species	Family	Leaf	Flower	Fruit	Other	Growth[a] form	Regeneration strategy[b]
43	*Faramea occidentalis*	Rubiaceae	x				t	s
44	*Ficus obtusifolia*	Moraceae	x		x	Stipules	t-he	s
45	*Ficus* sp.	Moraceae	x				he	
46	*Ficus tonduzii*	Moraceae	x				t	s
47	*Ficus yoponensis*	Moraceae	x		x	Stipules	t	s
48	*Gouania lupuloides*	Rhamnaceae	x				l	
49	*Guatteria dumetorum*	Annonaceae	x				t	s
50	*Guazuma ulmifolia*	Sterculiaceae	x				t	p
51	*Gustavia superba*	Lecythidaceae	x	x			t	i
52	*Hasseltia floribunda*	Flacourtiaceae	x				t	s
53	*Heisteria concinna*	Olacaceae	x				t	s
54	*Hibiscus rosa-sinensis*	Malvaceae	x				s	p
55	*Hippocratea volubilis*	Hippocrataceae	x				l	
56	*Hirtella triandra*	Chrysobalanaceae	x				t	s
57	*Hura crepitans*	Euphorbiaceae	x			Yellow leaves	t	s
58	*Hybanthus prunifolius*	Violaceae	x				s	s
59	*Hyeronima laxiflora*	Euphorbiaceae	x			Bark	t	
60	*Inga pezizifera*	Leg.-Mimos.	x				t	s
61	*Inga sapindoides*	Leg.-Mimos.	x				t	i
62	*Jacaranda copaia*	Bignoniaceae	x	x			t	p
63	*Lacmellea panamensis*	Apocynaceae	x				t	s
64	*Machaerium kegelii*	Leg.-Pap.	x				l-t	
65	*Machaerium riparium*	Leg.-Pap.	x				l-s	
66	*Macrocnemum glabrescens*	Rubiaceae	x	x			t	s
67	*Mangifera indica*	Anacardiaceae	x				t	
68	*Maripa panamensis*	Convolvulaceae	x				l	
69	*Miconia argentea*	Melastomataceae	x		x		t	p
70	*Mikania leiostachya*	Asteraceae	x				l	
71	*Mikania micranthea*	Asteraceae	x				l	
72	*Mikania tonduzii*	Asteraceae	x				l	
73	*Myriocarpa yzabalensis*	Urticaceae	x				s/t	
74	*Nectandra purpurascens*	Lauraceae	x				t	s
75	*Odontocarya tamoides*	Menispermaceae	x				l	
76	*Olmedia aspera*	Moraceae	x				s/t	s
77	*Paullinia baileyi*	Sapindaceae	x				l	
78	*Paullinia bracteosa*	Sapindaceae	x				l	
79	*Paullinia pterocarpa*	Sapindaceae	x				l	
80	*Petrea aspera*	Verbenaceae	x	x			l	
81	*Philodendron* sp.	Araceae	x				l/he (m)	
82	*Platymiscium pinnatum*	Leg.-Pap.		x			t	s
83	*Poulsenia armata*	Moraceae	x				t	s
84	*Protium costaricense*	Burseraceae	x				t	s
85	*Protium panamense*	Burseraceae	x				t	s
86	*Protium tenuifolium*	Burseraceae	x				t	s
87	*Pseudobombax septenatum*	Bombacaceae	x	x			t	p

Table 15. (*Continued*I

No.	Species	Family	Resource type				Growth[a] form	Regen- ration strategy[b]
			Leaf	Flower	Fruit	Other		
88	*Psychotria horizontalis*	Rubiaceae	x				s	s
89	*Pterocarpus rohrii*	Leg.-Pap.	x	x			t	s
90	*Quararibea asterolepis*	Bombacaceae	x	x	x		t	s
91	*Rheedia acuminata*	Guttiferae (=Clusiaceae)	x				t	s
92	*Serjania* cf. *decapleura*	Sapindaceae	x				l	
93	*Sloania terniflora*	Eleocarpaceae	x				t	s
94	*Sorocea affinis*	Moraceae	x				t	s
95	*Souroubea sympetala*	Marcgraviaceae	x				he	
96	*Spondias mombin*	Anacardiaceae	x				t	p
97	*Stigmaphyllon hypargyreum*	Malphigiaceae	x				l	
98	*Swartzia simplex* var. *ochnacea*	Leg.-Caesalp.	x				t	s
99	*Swartzia* sp.	Leg.-Caesalp.	x				t	
100	*Symphonia globulifera*	Guttiferae	x				t	s
101	*Tetragastris panamensis*	Burseraceae	x				t	s
102	*Theobroma cacao*	Sterculiaceae	x				t	s
103	*Tontolea richardii*	Hippocrataceae	x				l	
104	*Trattinickia aspera*	Burseraceae	x				t	p
105	*Trichilia tuberculata*	Meliaceae	x				t	s
106	*Trichospermum mexicanum*	Tiliaceae	x				t	p
107	*Trifolium* sp.	Leg.-Pap.	x				h	
108	*Triplaris cumingiana*	Polygonaceae	x				t	i
109	*Uncaria tomentosa*	Rubiaceae	x				l	p
110	*Unonopsis pittieri*	Annonaceae	x				t	s
111	*Virola* sp.	Myristicaceae	x				t	s
112	Unid. sp. 1		x					
113	Unid. sp. 2	Bignoniaceae	x				l	
114	Unid. sp. 3	Bignoniaceae	x				l	
115	Unid. sp. 4		x					
116	Unid. sp. 5	Leguminosae	x					
117	Unid. sp. 6	(cf. Moraceae)	x					
118	Unid. sp. 7		x					
119	Unid. sp. 8		x				l	
120	Unid. sp. 9		x				l	
121	Unid. sp. 10	Bignoniaceae	x				l	
122	Unid. sp. 11	Bignoniaceae	x				l	
123	Unid. sp. 12			x				
124	Unid. sp. 13			x				
125	Unid. sp. 14			x				
126	Unid. sp. 15		x					

[a] Growth forms: s = shrub; t = tree; h = herb; e = epiphyte; he = hemiepiphyte; l = liana; (m) = monocotyledon.

[b] Regeneration strategy: p = pioneer; s = shade-tolerant; i = intermediate.

Regeneration strategy: From the 72 species that could clearly be assigned to a particular regeneration strategy, 62.5 % belonged to the shade tolerants, 30.5 % to the pioneers and ca. 7 % to an intermediate category between pioneers and shade tolerants (according to Condit et al. 1996a). Since the proportion of these functional groups is a consequence of the species composition of the surrounding forest, we compared plants of the study plot (expected frequency) with host plants harvested along trunk trail III (observed frequency). Observed frequencies were significantly different from the expected (Table 16). Particularly the proportion of pioneer species was higher in the leaf-cutting ant harvest than in the forest. The observed frequency of cutting events (Fig. 46) also confirms this clear preference for light-demanding species.

Resource types: Green leaves were harvested from 96.8 %, flowers from 12.7 %, fruits from 6.3 % and other plant parts (e.g., nongreen stipules or bark) from 3.9 % of the visited species.

The spectrum of resources and food items selected by *A. colombica* in the present study is well in accord with general patterns described by published literature (reviewed by Vasconcelos and Fowler 1990; Farji-Brener 2001). Nevertheless, the following new observations need to be discussed. A preference for arborescent growth forms is known to be characteristic of many *Atta* ants (e.g., Blanton and Ewel 1985), while the proportion of lianas among the foraged species had never been exactly quantified before. Lianas account for ca. 13 % of the native species (Croat 1978) and approximately 25 % of the foliage on BCI (Putz 1988), and the 30 % lianas we found among the foraged species possibly indicates a disproportionate preference for this growth form. Such a preference could be explained by the fact that many vines and lianas, by showing high leaf production and turnover rates (Holbrook and Putz 1996), resemble pioneer trees that are generally known to be more susceptible to herbivores than shade-tolerant tree species (Feeny 1976; Rhoades and Cates 1976; Coley 1980, 1988). Similarly, in accord with current theories on general patterns in plant secondary metabolism and its role in defense against herbivores (Bryant et al. 1983; Coley et al. 1985; Coley and Aide 1991; Herms and Mattson

Table 16. Frequencies of plant regeneration strategies within the study plot and in the leaf-cutting ant harvest observed along trunk trail III. Classification according to Condit et al. (1996a), R. Condit, S.P. Hubbell, P.D. Coley (pers. comm.) and Dalling et al. (1998)

Regeneration strategy	Study plot[a]	Host plants along trail III[a]
Shade-tolerant	46	21
Pioneers	10	15
Intermediate	5	2

Goodness-of-fit: $\chi^2 = 17.88$; df=2; $p < 0.0001$

[a] Information on regeneration strategy was not available for eight plot species and five host species.

Fig. 46. Frequency of days with foraging observations on 52 host species throughout 1 year. Only species harvested along trunk trail III were included. Individuals of the same species were summed up. Total number of observation days was 126

1992), the high proportion of light-demanding tree species in the harvest of colony I implies that *Atta colombica* is a typical generalist herbivore discriminating against shade-tolerant species that possess a higher optimal defense level than fast growing pioneers (e.g., Coley 1982). These conclusions are supported by a recent paper of Farji-Brener (2001) who demonstrated by means of literature review and field assays that leaf-cutting ants generally prefer pioneers to shade-tolerant species. The present results on host tree selection also suggest a strong preference for large and mature individuals with thick trunks and voluminous canopies underpinning the view that *Atta* is a K-strategist (especially as compared with *Acromyrmex*, cf. Cherrett 1983) foraging mainly on nonephemeral and apparent plant resources. As dbh and tree height are positively correlated (e.g., Kira 1978) the favored harvest from thick-trunked trees means that herbivory of *A. colombica* is concentrated mainly in the topmost stories of the canopy. Using rope-climbing techniques, binocular observations, and the STRI canopy crane we could confirm this on several occasions (Weigelt 1996; pers. observ.). Similarly, Cherrett (1968) found *A. cephalotes* "allocating most of it's foraging effort in the high canopy" of a wet lowland forest in Guyana.

10.2.2 Harvest Diversity and Selection

Along trunk trail III, colony I harvested material from 72 trees consisting of 43 species (Table 15). Using the reciprocal of Simpson's index, harvest diversity was 30.14, which equals the number of equally common species required to generate the observed heterogeneity (Table 17). Evenness E calculated from the same index was relatively low for the total number of plant species harvested, but increased significantly to a maximum homogeneity of species abundance as the diameter of host trees augmented to >40 cm (Table 17; $r^2=0.94$; $p=0.0048$; $n=6$).

   The 0.22-ha study plot covers large parts of the area foraged from trunk trail III and therefore, permits a comparison between available and foraged resources. About 50 % of the species in the surrounding forest stand have been exploited during the observation period and this proportion did not change significantly with stem diameter, i.e., age structure (Fig. 47A; $r^2=-0.70$, $p=0.188$, $n=5$). There was, however, a remarkable tendency to cut leaves from tree individuals bigger than 20 cm dbh (Fig. 47B). This harvest preference could also be expressed by the electivity index, which ranged from −0.35 for stems between 5 and 10 cm dbh and 0.329 for stem diameters of 20–30 cm.

   Figure 46 ranks the plant species visited by the ants according to the frequency of harvest events per species throughout the course of the 126 monitoring days. As field notes on the intensity of fragment removal indicated, the frequency of harvesting is, to some extent, a measure of the amount of harvested biomass of each host tree. Therefore, quite obviously, the harvest is

Table 17. Characterization of the harvest diversity of colony I based on the plants collected along trunk trail III during 1 year of observation

	No. of individuals	No. of Species	Inverse of Simpson's Index	Evenness
Total plants	85	52	35.24	0.25
Total trees	72	43	30.14	0.25
Trees >5 cm	52	36	27.59	0.36
Trees >10 cm	41	31	25.08	0.46
Trees >20 cm	25	19	26.04	0.41
Trees >30 cm	12	10	9.00	0.62
Trees >40 cm	5	5	5.00	1.00

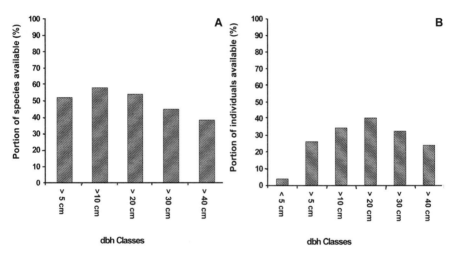

Fig. 47. Proportion of species (**A**) and individual plants (**B**) of different stem diameters selected by the ants from the available resource pool via trunk trail III

dominated by few highly preferred resources. The two most frequently harvested species were visited by the ants during more than half of all observation days. The first seven plants were exploited just as much (51 % of all foraging days) as all other 45 species together. Most of the latter species were harvested only once or for a few days during the year. The proportion of foraged species from the total number of available plant species has been used in many leaf-cutting ant publications as a principal measure for the degree of polyphagy. The present result of about 50 % corresponds well with the values in a review by Vasconcelos and Fowler (1990). However, this parameter may be misleading for an evaluation of leaf-cutting ant substrate selection since it does not consider the amount of material harvested from each species. The analysis of the harvesting frequency during this study (Fig. 46) suggests that the majority of the plant species gathered in the course of the year scarcely affects the quantitative assessment of the substrate diversity. In other words, as far as the quantities of the substrate species are concerned, *A. colombica* ants are rather oligophagous than polyphagous (see also Vasconcelos and Fowler 1990). This is further mirrored by the index for even distribution (evenness *E*) of resource species. There is a remarkable similarity with Shepherd's (1985) findings for a 12-month study on *Atta colombica* in a mature wet forest in Colombia (*E*=0.24 versus 0.25 in the present study), suggesting equal heterogeneity of resource exploitation in both studies. The harvest diversity of leaf-cutting ants has been shown to be affected by the floristic diversity of the habitat and the abundance of high ranking plant species respectively (for a review see Vasconcelos and Fowler 1990). The more evenly the resources are distributed (as in a mature tropical forest), the more species are included in the ants harvest. In contrast, in plant communities with lower diversity and predominance of a few attractive fungal substrates (e.g., pastures or mixed plantations), the ants' harvest will have a lower evenness index due to a few dominant species (cf. Blanton and Ewel 1985; Shepherd 1985).

Still, the large number of visited species compared to the few species accounting for the bulk of the harvest requires an explanation. The present observations show that samples are taken from many species, although the amount cut, but often not even carried away, is at many times negligible (pers. observ.). It is hard to believe that this rather low fraction of the total harvest does play a major role for colony fitness. Further, it seems unlikely that the heterogeneity of leaf types is needed to provide trace elements for optimal growth of the fungal symbiont (cf. Rockwood 1976). Although information on the nutrient requirements of the fungus is still scarce, this is indicated by the frequent and successful maintenance of *Atta* colonies in the laboratory with nothing but blackberry (*Rubus fruticosus*) or lilac (*Syringa vulgaris*). Therefore, it can be assumed that the diversity of food items collected by leaf-cutting ants is rather a consequence of the heterogeneous distribution of resources in time and space. Since host plant selection is strongly constrained by secondary plant metabolites (e.g., Cherrett 1972a,b; Howard and Wiemer 1986; Howard 1987, 1988; Nichols-Orians and Schultz 1990) which have been repeatedly shown to seasonally vary (Hubbell et al. 1984; Howard 1987), resource availability for leaf-cutting ant colonies is spatially and temporally restricted. To date these aspects of leaf-cutting ant foraging behavior are best explained by suggestions of Shepherd (1982, 1985) and Pyke (1984), who hypothesized that optimal foraging is achieved by constantly taking samples of all potential leaf sources, thus tracking variable distributions of substrate patches in time and space. However, it should be added that this monitoring behavior is additionally constrained by the nature of trail-centered foraging as discussed above (see Chap. 9.2).

11 Herbivory Rates

Before determining and assessing herbivory rates, it is crucial to first determine the objectives of the planned study (i.e., zoological, botanical or ecological focus) and also the level of ecological integration at which the measurements will be carried out because the parameters to be measured and the methods applied are highly dependent on this decision (Table 18).

Current research on plant-herbivore interactions shows a remarkable tendency to analyze herbivore effects at the level of the individual rather than at the population or community level or even at multiple levels (for review, see Zamora et al. 1999). In this chapter, we present a multiple-level analysis of leaf-cutting ant herbivory scaling all the way down from the whole island to the individual tree level.

11.1 Factors Affecting Estimates of Herbivory

The most common approach to quantify herbivory of single plants or plant communities is to determine the proportion of foliage removed by the herbivore. Recent research revealed that results of one-time leaf sampling from litter traps (Leigh and Windsor 1982) or plant canopies (Newbery and De Foresta 1985; Bongers et al. 1990) tend to markedly (up to five times) underestimate herbivory rates compared to long-term monitoring of tagged leaves (for reviews, see Landsberg and Ohmart 1989; Coley and Barone 1996). The latter method is now widely accepted as most appropriate since defoliation can accurately be expressed as a rate (Coley 1982; Lowman 1984; Filip et al. 1995). However, even this method has problems to estimate the effects of a single herbivore species since damage traces can rarely be identified properly. Further, published estimates of herbivory rates rarely refer to the ecosystem scale and due to variable sampling designs their interpretation is rather difficult (Lowman 1995).

Generally, for the collection and interpretation of herbivory data in tropical forests (with special reference to leaf-cutting ants), the following problems should be considered:

Table 18. Quantifying herbivory in ecological studies: the importance of the study level for the parameters to be measured

Study level	Objectives of study	Parameters to be measured
Primary consumer	– Ecology of the herbivore (with regard to, e.g., energy balance, optimal foraging, etc.)	– Quantity of plant biomass consumed – Temporal and spatial variation of harvesting pattern
Individual plant	– Ecology of the individual plant (e.g., effects of folivory on photosynthetic activity, water-use efficiency, growth rates, fitness, etc.)	– Leaf losses of the plant individual – Temporal and spatial variation of harvesting pattern
Community and ecosystem	– Acquisition of system-relevant baseline data (e.g., matter and energy fluxes) – Effects of herbivory (e.g., productivity, resource availability, competitive balance, species diversity, etc.)	– Consumption rates of herbivore taxa – Available biomass present in the system (e.g., net productivity or standing crop) – Relative loss of biomass from the system – Densities of herbivore and plant populations – Temporal and spatial variation of harvesting pattern

– *Temporal heterogeneity*
  Herbivory in seasonal tropical forests has been reported to be more intense in the wet season due to seasonal variations in herbivore abundance and food quality (Wolda 1978; Filip et al. 1995; Coley and Barone 1996; Gombauld and Rankin-de Merona 1998). Reduced dry season foraging on green leaves has also been shown for leaf-cutting ants (Rockwood 1975; Wirth et al. 1997). Consequently, estimates of annual folivory based on only wet season or only dry season data will reveal considerable over- or underestimations of herbivory rates (e.g., Lugo et al. 1973).

– *Spatial heterogeneity*
  Herbivory can be assumed to be lower in the closed forest than in tree-fall gaps, where typically less defended and fast-growing pioneer species accumulate (Coley 1982). Similarly, there is some evidence that leaf damage levels in the canopy are lower than in the understory (Lowman 1985; Barone 1994; but see Basset et al. 1992, who found higher arthropod densities in the canopy). Sampling designs with nonproportional representation of habitat types may therefore be misleading. Since the bulk of forest herbivory studies is naturally restricted to leaves accessible from the ground (Coley 1982,

1983b; Newbery and Foresta 1985; De la Cruz and Dirzo 1987) the resulting herbivory rates in forests may be prone to overestimation.

– *Size of foraging area*
For referring the consumption rates of a herbivore to the standing crop of the concerned vegetation, the size of the foraging area has to be known. As shown in Chapter 9, published estimates typically tend to overestimate the foraging area (Lugo et al. 1973; Haines 1978; Rockwood and Hubbel 1987). Consequently, the impact of herbivory may be underestimated.

– *Herbivore species*
Not all herbivores leave obvious evidence of their consumption and sometimes it is difficult to assess their impact. For instance, current estimations of total herbivory do not include the probably substantial losses caused by phloem feeders. Leigh (1999) suggested that a forest can loose as much or more dry matter to sucking than to chewing herbivores.

– *Damage type*
Leaf-cutting ants, like many other herbivores (e.g., vertebrates), often collect entire leaves (pers. observ.). The calculation of herbivory losses from one-time leaf samples naturally neglects these leaves, resulting in an underestimation of the total consumed leaf area (Odum and Ruiz-Reyes 1970; Leigh and Windsor 1982; Wint 1983; Newbery and De Foresta 1985; Bongers et al. 1990). Similarly, severely damaged leaves frequently wilt, die, receive posterior damage by pathogens (cf. Marquis and Miller Alexander 1992) and fall off, thus becoming excluded from sampling (Landsberg and Ohmart 1989).

– *Compensatory growth*
If compensatory growth plays a role (see Sect. 12.3), herbivory rates may be underestimated if harvesting rates and standing foliage area are not measured at the same time.

– *Restricted foraging efficiency*
Leaf-cutting ants tend to lose fragments during the transport along the extended foraging trails. Lugo et al. (1973) suggested that the yield at the nest entrances is about 70 % of the initially cut fragments. Since their activity is usually measured at the trail entrances (as in the present study), it can be assumed that the actual foliage removal is underestimated.

– *Leaf age*
Like other herbivores leaf-cutting ants seem to prefer young, not fully developed leaves (Rockwood 1976; Coley 1982; Waller 1982; Nichols-Orians and Schultz 1990). Thus the harvested foliage area does not necessarily reflect the entire loss of potential photosynthetically active leaf area.

Due to all these difficulties, published records of herbivory rates should be interpreted and generalized with care (cf. Landsberg and Ohmart 1989; Coley and Barone 1996). Because of the spatial and temporal dynamics of herbivoral damage as well as the peculiarities of food choice, the damage type and the foraging behavior of a broad spectrum of herbivores, only long-term studies covering all habitat types of a forest will reveal reliable herbivory rates. In the present work, we used an approach which is independent of individual leaf sampling. By measuring consumption rates of leaf-cutting ants and referring them to the estimates of a standing leaf crop of the surrounding forest, we obtained consumer-specific herbivory rates that allow interpretations at all scales from the individual tree to the ecosystem. As the above summary indicates, factors leading to underestimation of the actual herbivory damage clearly predominate. Thus, the following results should be regarded as the minimum effect caused by the ants.

11.2 Herbivory Rates at the Landscape Level

The herbivory by *A. colombica* on the landscape scale, i.e., within the 100-ha study area where most of the colonies were located (see Chap. 7.1) was estimated by applying the refuse-dumping activity method described in Chapter 8.4.2. From July 1997 to October 1998, monthly counts of the refuse-dumping activity of all colonies were conducted to estimate the total amount of harvested leaf fragments. This number was then used to calculate the fragment consumption rate of the colonies (see Chap. 8.1.2). Refuse-dumping activity within the colony population was nearly constant during the observation period. During 14 observations, the ants dumped an average of 98.5 ± 15.1 SD refuse particles min^{-1} $colony^{-1}$. Using the regressions from Chapter 8.4.3, a total annual number of harvested fragments of 2.1×10^9 can be estimated for the entire 100-ha area. Combining this value with the average harvest composition (i.e., ratio between green vs. nongreen material) and the mean weight and area of all harvested fragments from Table 19 results in the herbivory rates shown in Table 20. Comparing published values of leaf-cutting ant herbivory in tropical forests reveals a considerable variation of numbers (Table 21). Most studies so far have concentrated on a few single colonies, i.e., on the amount of harvested biomass in a discrete foraging area (Lugo et al. 1973; Wirth 1996). Cherrett (1989) combined data from different studies (consumption rates of single colonies cited in Lugo et al. 1973) and colony densities from various other studies and calculated a landscape level harvest rate of 0.253 g m^{-2} day^{-1} which would correspond to 923 kg ha^{-1} $year^{-1}$. However, being in the order of magnitude of the total herbivory estimated for neotropical forests (Leigh 1999), this value seems way to high just for leaf-cutting ants.

Table 19. Mean values of several characteristics of plant fragments harvested by randomly chosen *A. colombica* colonies throughout the 1997/1998 observation period on BCI

	Mean	±SD	n
Fraction (%) of:			
Leaf fragments among all harvested fragments	76.1	22.9	126
Fragments of young leaves among all leaf fragments	29.4	21.4	125
Flowers among all nonleaf fragments	36.1	39.6	119
Fruits among all nonleaf fragments	18.3	30.9	119
Ficus-stipules among all nonleaf fragments	34.3	40.6	119
Other fragments among all nonleaf fragments	11.4	20.7	119
Area (cm²) of:			
All leaf fragments	0.847	0.158	123
Fragments from young leaves	0.797	0.162	111
Fragments from mature leaves	0.881	0.156	120
Dry weight (mg) of			
All fragments	6.52	1.81	126
All leaf fragments	6.25	1.63	124
All fragments from young leaves	4.88	1.28	112
All fragments of mature leaves	6.83	1.98	122
All nonleaf fragments	7.41	3.98	106
Flowers	5.63	3.47	46
Fruits	7.81	6.40	31
Ficus-stipules	8.30	3.58	57
Other fragments	9.33	5.30	21

Table 20. Herbivory of *A. colombica* at the landscape scale determined for the entire 100-ha study area on BCI (Calculations based on the values for average harvest composition and the means for weight and area of all harvested fragments from Table 19)

Parameter	Amount	Unit
Daily loss of foliage area	3.64	$(cm^2\ m^{-2}\ day^{-1})$
Annual loss of foliage area	1330	$(m^2\ ha^{-1}\ year^{-1})$
Daily loss of biomass[a]	0.0368	$(g \times m^{-2} \times day^{-1})$
Annual loss of biomass[a]	134.4	$(kg\ ha^{-1}\ year^{-1})$
Herbivory rate[b]	2.5	(%)

[a] Including nongreen material.
[b] Based on a leaf area index of 5.25 (see Chap. 6).

Table 21. Published harvest rates of leaf-cutting ants

Harvest rate (kg biomass ha⁻¹ year⁻¹)	Habitat	Number of studied colonies	Duration of study	Source
A. cephalotes:				
653	Experimental plots	3–4	5 months	Blanton and Ewel (1985)
A. colombica:				
517	Secondary moist forest	1	1 year	Wirth (1996)
297	Wet forest	1	8 days	Lugo et al. (1973)
98	Secondary moist forest	9	4 months	Haines (1978)
134	Secondary moist forest	Approx. 50	16 months	This study

Herbivory rates are frequently expressed as the consumed proportion of the standing crop. The problems with direct determinations from marked leaves in the canopy or shed leaves on the ground have already been mentioned above. To overcome these difficulties, in the present study, the standing leaf crop was determined by applying a light interception model for LAI estimation (see Chap. 6). Using the calculated average LAI of 5.25 (see Table 6) and the average fraction of leaves within the total harvest of *A. colombica* (Table 19), the landscape level herbivory rate of these ants amounts to approx. 2.5 %. This value is roughly in the order of magnitude of the 1.6 % in the study of Haines (1978) carried out in another Panamanian forest close to BCI, but is clearly lower than the 17 % which can be taken from the above-mentioned study of Cherrett (1989). Since overall herbivory in neotropical forests typically lies around 11 % (Coley and Barone 1996; see also Tables 22 and 23), *A. colombica* would contribute approximately one fifth to one sixth of the total. This proportion is much lower than the commonly quoted 80 % of apparent leaf damage caused by leaf-cutting ants published by Wint (1983). This high value, however, is not based on a systematic study, but rather reflects the author's impression from a "casual observation", it may still be right on a scale smaller than the individual tree level (see below).

Nevertheless, although smaller than previously assumed, leaf-cutting ants certainly do harvest a significant proportion of the total consumption of matter and energy in the landscapes where they occur. Up to now, it is unclear whether the leaf consumption of leaf-cutting ants is taken over by other herbivores in areas where the ants are lacking. Thus, it remains open, whether the herbivory by other consumers is reduced when the ants are present and whether this would affect the productivity and the number of higher trophic levels (McNaughton et al. 1989). Interestingly, leaf-cutting ants may in fact play a special role in trophic interactions, since predation on adult colonies seems to be very low (Rao 2000) and therefore, the consumed energy and mat-

Table 22. Important species and groups of herbivores on BCI and their herbivory and folivory rate. (after Leigh 1999, and own results)

Species/group	Herbivory rate (kg ha^{-1} year^{-1})	Folivory rate (kg ha^{-1} year^{-1})
Alouatta palliata (Howler monkey)	88	35
Bradypus variegatus (three-toed-sloth)	110	110
Choloepus hoffmannii (two-toed-sloth)	22	22
Mazama americana (brocked deer)	$\leqq 60$	?
Tapirus bairdii (Baird's tapir)	14	$\leqq 14$
Coendou rothschildii (coendou)	9–95	9–95
Iguana iguana (iguana)	7	7
Folivorous insects	476–624	476–624
Atta colombica[a]	134.4	102.3

[a] Present study.

Table 23. Selected literature values for total annual folivory in tropical forests and on BCI

Folivory rate (%)	Forest type	Method	Source
11.1	Moist forests – late successional	Meta-analysis of 21 studies	Coley and Barone (1996)
48.0	Moist forests, pioneer species	Meta-analysis of 4 studies	Coley and Barone (1996)
14.2	Dry forests	Meta-analysis of 4 studies	Coley and Barone (1996)
7.24	Secondary moist forest (BCI)	Determination of hole sizes in shed leaves	Leigh (1999)
14.6	Secondary moist forests (BCI), tree-fall gaps	Monitoring of marked leaves from late successional species	Coley (1983b)
8.3	Crown area of a dry forest near Panama City	Monitoring of marked leaves from 9 species	Leigh (1999)

ter gets channeled through a "trophic shortcut" to decomposition (cf. Chap. 14). On the other hand, it could also be that the trophic influence of leaf-cutting ants runs relatively uncoupled from other trophic cascades. Knowing that leaf-cutting ants obtain the greatest amount of their harvest from the higher canopy strata (see Chap. 10), and considering the fact that major parts of the total herbivory of the system take place in the understory (Basset et al. 1992; Lowman 1995; Coley and Barone 1996), it may well be that the understory herbivores are more or less unaffected by the leaf-cutting ants. Unfortunately, information on possible effects of competitive exclusion on trophic interactions (as reported for systems with grass-cutting ants and grazing cattle by Fowler and Saes 1986 and Fowler et al. 1986a) is presently not available for forests.

11.3 Herbivory at the Colony Level

In this section, we try to realistically estimate the actual foliage loss within the foraging area of colony I. Naturally, such an estimation depends on the definition of the foraging area. In order to demonstrate the importance of this parameter, we present several estimates always using the same harvesting data, but varying the degree of precision in determining the foraging area. Trying to link the observed removal rates as closely as possible to the standing leaf crop, the analysis was restricted to the harvesting activity along foraging trail III which was largely situated within the study plot where the LAI data were sampled (see Chap. 6).

11.3.1 Leaf Harvesting Rates Along Trunk Trail III

Figure 48 shows the variation of the leaf harvest activity along foraging trail III and the total biomass input to colony I throughout the year (see Chap. 8). While the daily inputs from trail III basically follow the seasonal harvesting pattern of the whole colony, the relative contributions to the total nest input varied throughout the year. For example, ant activity virtually ceased in the dry season, but accounted for up to 30 % of total colony activity in the wet season (Table 24). Nevertheless, the total annual leaf area input estimated for foraging trail III in Table 24 suggests that this sector III can be considered as representative for the colony. Breaking down the total area of the leaf harvest per year (3855 m^2) in proportion to the relative sizes of the four foraging sectors (cf. Fig. 45C) reveals that the actually counted harvested leaf area from trail III, i.e., 993 m^2 year^{-1}, differs only slightly from the expected figure of 982 m^2 year^{-1} (i.e., approx. 25 % of the total). This implies that the ants of colony I not only partitioned their potential foraging range into four equally sized subareas, but also, on an annual basis, removed almost equal amounts of foliage from these subareas.

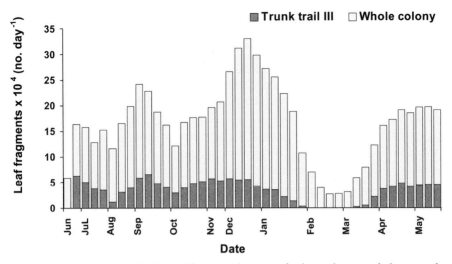

Fig. 48. Annual pattern of daily leaf fragment harvest of colony I harvested along trunk trail III and at the colony level (cf. Fig. 25) expressed as moving average over 4 days of observation. *Ticks* on the *abscissa* represent the beginnings of the labeled months (four values per month, except for November)

Table 24. Annual and seasonal harvest rates of the entire colony I and from foraging trail III ($n=45$, $n_{dry\ season}=17$, $n_{wet\ season}=28$)

	Total leaf area (m²)	
	Colony I	Trail III
Annual	3855	994
Dry season	1115	184
Wet season	2740	810

11.3.2 Defining a Discrete Foraging Area

In order to define the size of the foraging area as precisely as possible, the crown areas of the foraged trees were estimated during the course of the study year. Projected areas of the crowns of 12 selected host trees along trunk trail III were plotted on a map of the study plot by measuring the contours of the tree crowns. At least four edge points were located for each tree from the four main compass directions using the trunk of the tree as a reference point. If the horizontal shape of the crown was very irregular, the number of mapped edge points was increased. With the help of the estimated edge points, sketches of the projected horizontal shape of the respective tree crowns were drawn. Projected crown areas were then estimated by cutting out the drawn crown projections and determining their area with a leaf area meter (LI-3000, LI-COR, Lincoln, Nebraska, USA).

Comparisons of projected crown areas and the respective trunk diameters measured at breast height revealed a highly significant exponential relationship ($r^2=0.951$, $p<0.001$, $n=12$; Fig. 49). This regression equation was then used to obtain an approximation of the total crown area of all trees (no lianas) in the area of trail III which were foraged by the ants during June 1993 and June 1994. Figure 50 shows both the directly measured crown contours (polygons) and those estimated from the regression (circles) of all these host trees. The total enclosed area (FA_{discr}) was 1323.76 m^2. Two other areas, FA_1 and FA_2, represent estimates with a lower degree of precision. These estimates include different proportions of non-foraged forest area and their values are, therefore, higher than Fa_{discr} ($FA_1=2640$ m^2 and $FA_2=4800$ m^2).

11.3.3 Annual Foliage Loss from the Foraging Area

Knowing the size of the foraging area served by trail III allows the calculation of the total annual herbivory within this area (Table 25). Assuming that there is no compensatory growth, the stand loses 12.5% of its standing foliage within FA_{discr} Using the larger FA_1 and FA_2 instead of FA_{discr}, leads to values of 6.66 and 3.79%, respectively. Using the estimated ratio between leaf area and

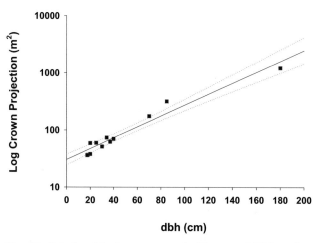

Fig. 49. Relationship between trunk diameter (dbh) and crown projection area for selected host tree individuals of the species *Anacardium excelsum, Apeiba membranacea, Guazuma ulmifolia, Hura crepitans, Inga pezizifera, Jacaranda copaia* (two individuals)*, Miconia argentea, Protium panamense, Protium tenuifolium,* and *Trattinickia aspera* [(log (area)=0.00955 dbh+1.48; $r^2=0.951$]. Dotted lines represent the 95% confidence interval

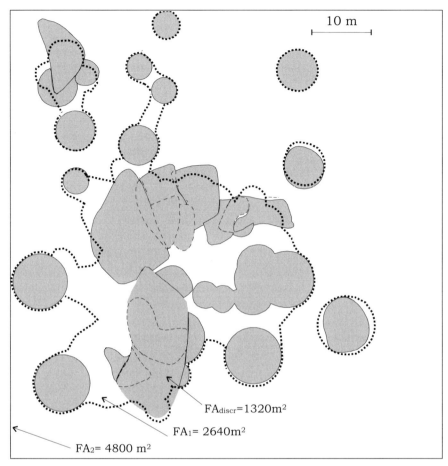

Fig. 50. Discrete foraging area of trunk trail III (FA$_{discr}$: *gray areas*) and two other estimates of foraging areas (FA$_1$: *dotted line*, FA$_2$: *rectangle*) with lower degrees of precision
(see text for further explications). The area calculation of FA$_1$ is based on the contours in
Fig. 44

Table 25. Calculation of the annual herbivory of colony I within the area of foraging
trail III

Estimated foraging area (FA$_{discr}$)	1323.76 m^2
Calculated mean LAI	5.25 m^2 m^{-2}
Total leaf area within FA$_{discr}$	6949.74 m^2
Total leaf area removed during the observation year	993.5 m^2
Percentage of removed relative to standing leaf area within FA$_{discr}$ (without compensatory growth): 993.5 m^2/(993.5 m^2+6949.74 m^2)=	12.50 %

biomass (cf. Chap. 8.2), these values correspond to a harvested biomass of 517 kg ha^{-1} year^{-1} for FA$_{discr}$, 259.4 kg ha^{-1} year^{-1} for FA$_1$ and 142.7 kg ha^{-1} year^{-1} for FA$_2$.

With regard to a reported overall herbivory of about 11 % for tropical shade-tolerant species in the understory (Coley and Barone 1996) and probably even lower values in the canopy (Barone 1994), the above proportion of about 13 % herbivory within the discrete foraging area of the colony seems to be way too high for a single herbivore species and the values for FA$_1$ and FA$_2$ seem much more realistic. This discrepancy nicely illustrates the problem that the mere radius of action based on the potential foraging range of a herbivore may be a rather unsatisfactory basis for judging its phytophagous relevance. In the case of semisessile (cf. Chap. 7) and trail-centered leaf-cutting ants, large parts of the area described by their radius of action remain unaffected (cf. Chap. 9), while other areas are heavily exploited. The little available information from only three publications relating harvesting rates to colony foraging area differ largely (Table 26). This might be due to the neglecting of seasonal variations in leaf harvesting (all authors measured activity only in the wet season), or to the difficulty of comparing activity counts with leaf input rates estimated from measurements of refuse disposal (Haines 1978; see Claver 1990 for discussion). Even the good agreement of the present estimate with Blanton and Ewel's (1985) data is likely to be coincidental as they quantified the overall consumption rate of several *Atta cephalotes* colonies feeding in plots of experimentally designed, 1.5-year-old plant communities. However, since calculated herbivory rates are basically a function of the precision of the foraging area determination, we believe that the differences between the values in Table 26 mostly illustrate the importance of how to define the corresponding foraging area. This situation can only be improved if more attention is paid to a precise determination of any herbivores' foraging area and to the development of compulsory standards. Generally, it is difficult to extrapolate from colony-based figures to the whole ecosystem since estimates of leaf-cutting ant activity are typically nest-specific and can, therefore, not

Table 26. Daily biomass harvest rates of leaf-cutting ants per square meter forest floor as reported by various authors

	Haines (1978)	Lugo et al. (1973)	Blanton and Ewel (1985)	Present study
Harvest rate (g m^{-2} day^{-1})	0.027	0.081	0.149	0.141
Method of foraging area determination	Study area (28 ha) equals foraging area	Rectangular contour of presumed foraging area	Use of experimental plant communities of defined size	Crown projection area of all host plants (discrete foraging area)

simply be converted to leaf consumption per unit ecosystem area (cf. Fowler et al. 1990).

11.4 Herbivory at the Individual Plant Level

11.4.1 Methodological Details

Definition of harvesting periods: Information from the monitoring of foraging trails (see Chap. 9.1) was used to estimate the number of harvesting days per year for selected host trees. Since a single host plant could be foraged several times throughout the year, harvesting periods for each observed host tree were estimated. Harvesting activity at a particular host tree was interpolated between two consecutive observation days, which were typically at intervals of 2–3 days. In contrast, a single harvesting period was assumed to be completed if cutting activity was not resumed on a consecutive observation day.

Estimation of leaf harvesting rates per harvesting period: Daily harvest from a specific tree was estimated using the regression relationship presented in Chapter 8.4.1 and the counts of peak foraging activity obtained at the weekly census days (cf. Chap. 8.1). The total leaf area, LA_{tot} (m^2), removed from a tree individual during a harvesting period results from the equation:

$$LA_{tot} = \frac{D * N_{frag} * A_{frag}}{10000} \qquad (8)$$

where D is the number of foraging days per harvesting period, N is the mean estimated number of fragments per day, and A_{frag} (cm^2) is the mean fragment area for the respective species as measured from fragment samples of all observation days (cf. Chap. 8.1). Annual leaf area removal from a particular tree can then be calculated as the sum of all partial amounts during the harvesting periods. If no fragment count was performed during a harvesting period, the average input for the concerned tree was used instead.

11.4.2 Estimation of Annual Leaf Harvest
from Selected Individual Host Plants

Combining data from the trail monitoring (Chap. 9.1) and the study of foraging activity (Chap. 8.1) lead to an approximation of typical features of leaf-harvesting activity at the level of individual host plants of colony I. The

Table 27. Features of leaf harvesting on selected host plants of colony I and estimation of annual leaf consumption

Leaf source	Harvesting frequency (day year^{-1})	Mean fragment area (cm^2)[a]	DLLA[b] (with LAI = 2,5 (m^2)	Total annual leaf harvest (m^2)[c]
Apeiba membranacea	90	0.65	438.37	33.67
Cecropia insignis	128	0.48	150.02	42.19
Guazuma ulmifolia	82	0.69	91.34	17.94
Hura crepitans	48	0.8	795.63	195.26
Stigmaphyllon hypargyreum	85	0.77	–	108.58
Trattinickia aspera	17	0.82	94.67	37.03

[a] Means derive from *n* combined samples containing a total of *N* leaf fragments as follows (*n, N*): *A. membranacea*: 14, 247; *C. insignis*: 19, 322; *G. ulmifolia*: 16, 166.5; *H. crepitans*: 13, 1134.5; *S. hypargyreum*: 13, 543; *T. aspera*: 4, 261
[b] For definition see footnote on page 146
[c] Estimates by summing up subtotals of single harvesting periods calculated by means of Eq. (8).

five tree and one liana species which were investigated in this connection can all be considered as highly ranked leaf sources (cf. Fig. 46), although the estimated total amount of leaf consumption per year revealed a great variability (Table 27). Further, the influence of species-specific leaf properties on the quantity of leaf removal could clearly be demonstrated. For example, harvesting frequency on the "hard-to-cut-leaves" of *Cecropia* was almost three times as on the soft-leaved *Hura,* but the harvested *Cecropia* leaf area was less than one fourth compared to the harvested leaf area from *Hura.* Similar effects can be observed on the counterparts *Apeiba* and *Trattinickia* (Table 27).

11.4.3 Annual Pattern of Leaf Consumption and Herbivory Rates

Analyzing the temporal pattern of leaf harvest throughout the year provides a more detailed view of individual plant usage (Fig. 51). Generally, the extent and the duration to which the different resources were used seem to be roughly in accord with the seasonal patterns described for colony I in Chapter 8 (cf. Fig. 25). This is particularly true for the pronounced decrease during the dry season between January and April. In contrast, the time of maximum leaf harvest during the rainy season seems to be rather species-specific, probably indicating species-dependent seasonal changes of leaf quality. For example, whereas *Apeiba* was cut mostly in May, the foraging peak for *Hura* fell in November, and that for *Trattinickia* in June. However, it is important to point out that harvesting on all of the six species did not occur during a unique period, but was resumed several times throughout the year.

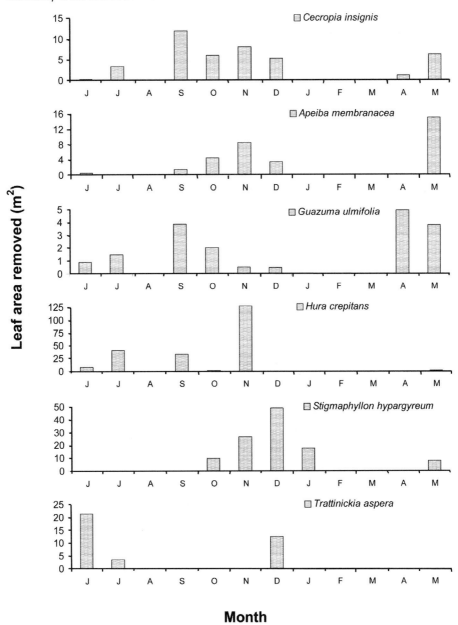

Fig. 51. Annual pattern of leaf area removal by ants of colony I from selected host plant individuals

Based on the vertical LAI distribution measured in the forest (see Fig. 19), DLLA* values were estimated by taking the measured and estimated crown projection areas of the five trees (see Fig. 49; Table 27). Since all the trunks were thicker than 20 cm dbh, it was assumed that their crowns were mainly in the top canopy layer (cf. Hallé et al. 1978). Hence, according to Fig. 19, a mean LAI of 2.5 seems to be a reasonable value to assess DLLA for these trees (Table 27).

The resulting fractions of leaf removal shown in Fig. 52 represent approximations of leaf-cutting herbivory on preferred host plants. Foliage losses ranged widely from less than 10 % to almost 40 % of the total leaf area of an individual tree crown per year. However, considering the variability of the temporal distribution of these losses (Fig. 51) it appears that high amounts of harvested foliage do not necessarily correspond to a long duration of harvesting as can be seen in the case of *T. aspera*, which lost 40 % of its foliage in just 17 days. Thus, depending on the particular species, the same annual leaf loss can be a result of rather different temporal harvesting patterns ranging from a few days during 1 month (e.g., ca. 25 % in *H. crepitans*) to many days during several months (e.g., ca. 28 % in *C. insignis*; see Table 27 and Fig. 51).

Due to the relatively high variability of the daily input rates taken at weekly intervals, estimations of herbivory rates at the single tree level will only yield

---

* The DLLA (Drip Line Leaf Area) is an leaf area related to the area covered by single crowns and can be used as a measure of total leaf area of a plant individual. It is calculated as the product of projected crown area and LAI.

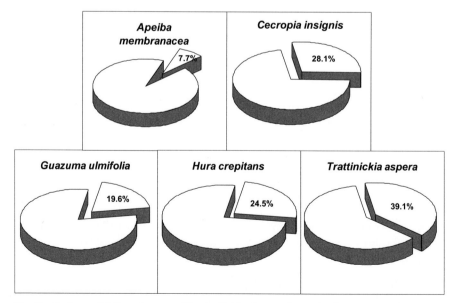

Fig. 52. Estimated foliage losses of five individual trees due to leaf-cutting ant herbivory

approximate results. Nevertheless, the values between 8 and 40 % for the five selected tree species give at least some idea of the potential magnitude of defoliation of plant individuals by leaf-cutting ants in natural habitats. With the exception of some anecdotal reports of total defoliation (Rockwood 1975; Marquis and Braker 1994) estimations of single plant damage rates due to *Atta* species have not been published. This is remarkable since the observation of partial defoliation (Cherrett 1968; Fowler and Stiles 1980) gave rise to so many discussions about the foraging strategies of leaf-cutting ants (see Chap. 9). One must bear in mind though that values higher than 20 %, as given in the present study, certainly represent the high end of leaf-cutting ant herbivory since higher ranked host plants of the studied colony were intentionally selected for this investigation. Still, even as a potential grazing rate, these figures are interesting because they are higher than the literature values for overall herbivory (other than outbreaks) in tropical plants. The respective studies were performed by Dirzo (1984b) in Mexico (0–31 %), Bongers et al. (1990, 0.8–11 %) and Sterck et al. (1992) in French Guyana (1.6–12.8 and 0.8–9.3 %), Coley (1983b) with mature leaves of tree saplings in Panama (11±51 % SD and 44±91.5 % SD), or Marquis (1984) with *Piper arieianum* in Costa Rica (4.3–30 %), most of the authors emphasizing the extremely high variation between individuals. The spatially restricted foraging behavior of leaf-cutting ants may be interpreted as one possible cause for this variability. Distribution of leaf damage levels among plant individuals in a tropical forest is generally assumed to be highly unpredictable (Coley 1983b). However, the present results showing repeated cutting activity of attractive plant individuals along the long-term foraging trails of *Atta* colonies, provide strong evidence that this does not apply for leaf-cutting ant herbivory.

Although there is a controversial discussion about the selective impact of folivores (see Marquis 1992b for review), it has been shown that an annual leaf loss of only 10 % is sufficient to reduce plant fitness of tropical woody plants (Dirzo 1984a; Marquis 1984, 1992a). A relatively low overall damage level would play a subordinate role, whereas natural selection rather depends on variable damage levels around the mean.

Amongst many other variables affecting plant responses to herbivory, the importance of the timing of the damage relative to the timing of resource accumulation and environmental constraints on leaf phenology has been stressed by several authors (Marquis 1992b; Haukioja and Honkanen 1996; Crawley 1997). The annual pattern of leaf removal by *A. colombica* showed that not only leaf losses, but also the length and number of harvesting periods differed widely between the selected plants. Hence, depending on the phenological status of the concerned tree, this may influence compensatory refoliation (Tuomi et al. 1994), defense induction (Norris 1988), or induced alteration of the phenology of leaf production as a mechanism to avoid herbivory (Coley and Barone 1996). Up to now, there is only limited knowledge available about the consequences of leaf-cutting ant herbivory at the single plant level for the fitness of the individuals. Detailed comparative long-term studies with sufficient replication are needed to improve this situation.

11.5 Comparison of Herbivory at Different Levels

Removing sizeable amounts of foliage from the trees, leaf-cutting ants certainly play an important role among herbivores of comparable taxonomic diversity within the tropical rainforests of Central America. However, mere measurement of nest input rates as they are commonly reported in the literature (e.g., Lugo et al. 1973; Haines 1978) cannot adequately describe the impact of *Atta* herbivory on the plant community. Depending on the scale of the investigation, the percentage of foliage loss due to *Atta colombica* ranges from almost nothing up to ca. 40% (Fig. 53). Thus, the level of ecological integration of a study is of crucial importance for any future quantitative evalua-

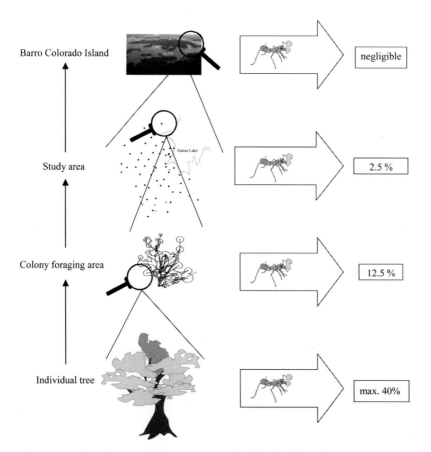

Fig. 53. Leaf-cutting ant folivory at various levels of ecological integration

tion of *Atta* herbivory on the tropical rainforest ecosystem. *Atta* ants are selective herbivores which create a high level of spatiotemporal heterogeneity in their habitat. For an ecologically meaningful analysis of the impact of these ants on their habitat it is, therefore, necessary to gain quantitative information on the following phenomena (for details see Wirth 1996; Herz 2001): (1) there are strong preferences for specific host plants; (2) individual tree crowns are typically defoliated in patches; (3) foraging territories are explored along discrete and persistent trunk routes; (4) forests are nonhomogeneously colonized by ant colonies; (5) daily and seasonal foraging patterns are affected by rainfall patterns and host plant phenology; (6) there is a rather high colony turnover rate (colony movements) with yet unknown causes.

12 Herbivory and Light

In Chapter 5, the light climate within the rainforest was assessed within the vicinity of a leaf-cutting ant colony. These measurements highlighted the enormously variable light conditions both vertically through the canopy and near the forest floor. Since the cutting of leaves by ants affects canopy structure and subsequent light penetration (Plates 12, 13), the significance of this foliage removal to the light climate and primary production of the canopy becomes an interesting issue. As mentioned earlier, light availability may limit growth and survival of many plant species by limiting photosynthesis, and heterogeneity in the light environment may be an important functional attribute of the rainforest ecosystem. This chapter focuses on the measurable consequences of leaf harvesting by ants on the canopy light climate and potential primary production.

12.1 Effects of Foliage Removal on Light Distribution and Intensity

12.1.1 Changes in the Understory Light Climate of the Foraging Area

The available data on the horizontal distribution of diffuse transmittance above the ground within the study plot (see Chap. 5) were examined for signs of changes in relative irradiance in the understory of the leaf-cutting ant foraging area. About 40% of all measurement grid points in the 2200-m² plot were contained within five variably sized foraging patches (Fig. 54). Although the mean transmittance inside the foraged areas was lower (1.8±1.3 SD, $n=155$) than outside (2.2±1.2 SD, $n=225$), the distribution of light differed significantly between the two forest sections (Kolmogoroff-Smirnoff: $p<0.005$). High light transmission (>5%) appeared primarily inside the foraged patches, and the mean of transmittance values >5% was higher in the foraged areas (T=2.46; df=13; $p<0.05$). This strongly suggests

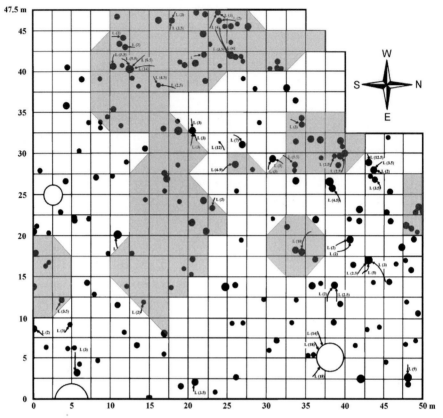

Fig. 54. Foraging areas (marked *gray*) of colony I within the grid system of the study plot. (cf. also Figs. 9 and 50)

that overall differences in light climate may be related to differences in patches of high light transmittance.

Overall, the light regime in the foraging area can be characterized as similar to the surrounding forest, but exhibits higher variability in light transmittance. Is this an effect of small foliage gaps caused by leaf-cutting activity at discrete canopy locations, or is it merely a consequence of nonhomogeneous forest structure in the plot? Assuming a close relationship between crown size and trunk diameter (see Fig. 49), we compared trunk diameter distributions between the two forest sectors, and found that the distribution of trunk diameters was similarly distributed both within and outside the foraging areas (Kolmogoroff-Smirnoff: *p*>0.1). This indicated that areas of high transmittance are probably due to leaf removal by ants rather than structural differences.

12.1.2 Changes in the Light Climate of an Individual Tree

Removal of foliage within a tree canopy by leaf-cutting ants reduces light interception by the affected foliage and permits light to pass through the resulting gaps to foliage below. The pattern of the light can affect photosynthesis rates of the foliage below (Pearcy et al. 1994; Ryel and Beyschlag 2000). To assess these changes in light, Weigelt (1996) measured light conditions in a *Pseudobombax septenatum* tree subjected to ant foraging. Fully grown *Pseudobombax septenatum* are overstory trees in the rainforest. They are also preferred foraging trees for *Atta*, and removal of foliage from these trees will affect the light climate of much of the foliage below.

The light microclimate within and below the single *Pseudobombax* tree was measured with a fixed array of light sensors (Fig. 55) before, during, and after foraging by leaf-cutting ants. Measurements were conducted over a 60-day period commencing after initiation of leaf production in this deciduous

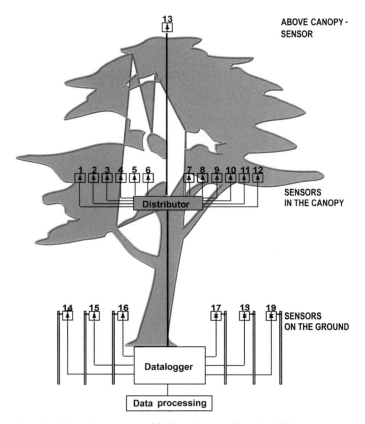

Fig. 55. Measuring array of light sensors placed within an overstory *Pseudobombax septenatum* tree subjected to leaf-cutting ant foraging. A single light sensor was also placed above the tree crown to measure light above the canopy

species. Two significant changes in light patterns due to ant foraging were observed from these measurements. First, the variability in light reaching a single point increased greatly (Fig. 56). This was measured as the change in percentage of time a light sensor was exposed to full sunlight (sunflecks) between consecutive days. Ant-induced gaps in foliage increased the variability of this value as sun penetrated through different 'holes' during different days. The second change was the increase in the frequency of sunflecks (short periods of full sunlight) reaching the sensors under the canopy where ants foraged. However, while the time increased when direct sunlight reached the sensors, the flux of diffuse light did not change appreciably with foliage removal (Fig. 57). These results indicate the major effect of foliage removal within this tree was the increase in the frequency and variability of sun flecks.

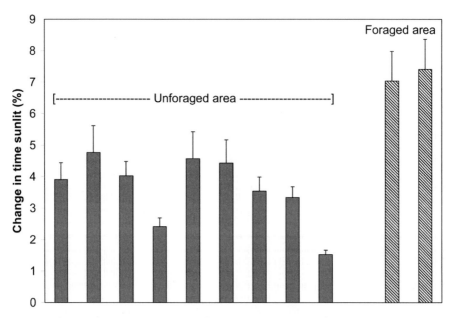

Fig. 56. Changes between consecutive days in the percentage of time that light sensors were illuminated by full sunlight. Each *bar* represents a single light sensor. Nine sensors were located in canopy regions where ants did not forage, while two sensors were located in regions where ants removed foliage. Sensors were located in the mid-canopy regions (see Fig. 55)

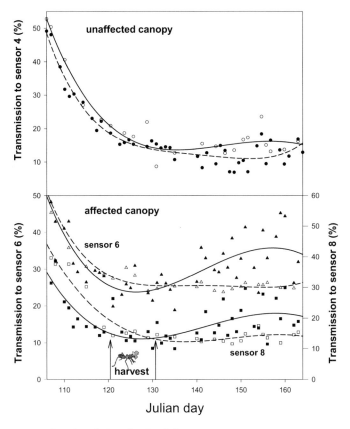

Fig. 57. Light transmission to three individual sensors within the canopy of a *Pseudobombax septenatum* tree where the light climate was unaffected (*above*) or affected (*below*) by foliage removal by leaf-cutting ants. Transmission of both diffuse and direct short wave radiation are shown. Diffuse radiation (*open symbols*) is expressed as the percent intensity of the above canopy sensor, while direct beam radiation (*closed symbols*) is expressed as the percent of the daylight period where the sensor was exposed to direct sunlight. *Lines* represent cubic polynomial fits using least-squares regression

12.1.3 Simulated Changes in Canopy Light Intensity

Leaf-cutting ants exhibit a particular preference for large trees (cf. Chap. 10, see also Cherrett 1968), and prefer sun leaves to shade leaves (Berish 1986; Nichols-Orians 1991a). From this and our own field observations, it can be concluded that foraging *Atta* workers will be most numerous in the upper forest strata and a major part of the total leaf harvest will take place in the topmost portion of the canopy. Since light extinction is exponential (cf. Fig. 16), the relative impact of leaf-cutting activity on light interception throughout the canopy will be greater when foraging occurs in the top canopy layers. This was demonstrated using a simple simulation of light extinction through the canopy with foliage removed in various portions of the canopy. We calculated light extinction throughout the canopy with 10 % of the total foliage removed in the upper 20 % of the canopy (as defined by cumulative LAI), from the middle 20 %, the lowest 20 % and uniformly throughout the canopy, and compared this with the light extinction for an intact canopy.

Light extinction was substantially less when foliage was removed from the upper 20 % of the canopy (Fig. 58) as compared to the canopy with no foliage removed. This was also the case when compared to canopies with foliage removed uniformly or from the middle or bottom of the canopy. In fact, light intensity was 35 % higher throughout the canopy with foliage removed from the upper portion of the canopy as compared to the unforaged canopy. Such increases in light were only seen in the lower half of the canopy with foliage removed from the middle canopy, and only from near the canopy bottom with foliage removed from the lower canopy (Fig. 58). The canopy with foliage removed uniformly had less than 10 % increase in light intensity in the upper third of the canopy as compared to the unforaged canopy, but this increased roughly linearly to 34 % at the bottom of the canopy.

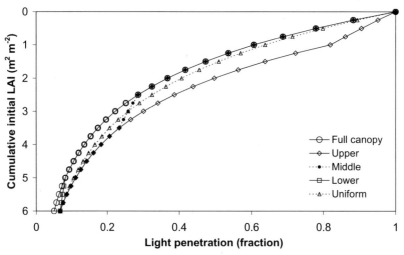

Fig. 58. Simulated instantaneous light penetration by canopy depth (expressed as cumulative LAI for full canopy) for a rainforest canopy with an initial total leaf-area index of 6-m² foliage area per m² ground area with different patterns of foliage removal. Values represent the fraction of diffuse radiation or area of sun flecks attenuating through the canopy. Simulations consider the removal of 10 % of the total foliage from: the upper 20 % of the canopy (*upper*), the middle 20 % of the canopy (*middle*), the lower 20 % of the canopy (*lower*), and uniformly from the whole canopy (*uniform*). Also shown is light penetration with no foliage removed (*full canopy*)

12.2 Canopy Photosynthesis

12.2.1 Simulated Changes in Whole Canopy Photosynthesis

Patterns of foliage removal affect not only the canopy light climate, but also whole canopy photosynthesis rates. Removed foliage reduces the leaf surface area where photosynthesis can occur, but increased incident light on the remaining foliage can result in higher photosynthesis rates. Perhaps higher photosynthesis rates on the remaining foliage can compensate in part for the loss of foliage. Reductions in whole canopy net photosynthesis could reduce community primary production and have significant implications on ecosystem energy flow and whole ecosystem biomass.

We explored the effects of foliage removal by leaf-cutting ants on whole canopy photosynthesis using a simulation model. The model used consists of two primary components: a submodel to simulate environmental conditions within the canopy (Caldwell et al. 1986; Ryel et al. 1990, 2001), and a submodel to calculate single-leaf photosynthesis based on leaf physiological properties and the environmental conditions to which leaves are exposed (Farquhar et al. 1980; Farquhar and von Caemmerer 1982; Harley et al. 1986). For simulating environmental conditions within the canopy, the canopy is considered to be composed of layers of relatively uniform foliage density and orientation. The submodel uses this structure to then simulate the penetration of light through the foliage layers and also the profile of microclimatic conditions such as temperature, wind and humidity. For foliage within the canopy layers, photosynthesis rates are calculated for sunlit and shaded leaves by the photosynthesis submodel based on the simulated environmental conditions. Whole canopy photosynthesis is then calculated by weighted averages of sunlit and shaded foliage within canopy layers, and the distribution of canopy foliage among layers. Values are expressed per unit ground area. The model obviously represents a significant simplification of a system as complex as a tropical rainforest, but sizeable effects of foliage removal on canopy photosynthesis can be effectively assessed. This modeling approach has been used to address a wide range of ecological questions in a variety of plant communities (e.g., Harley et al. 1986; Grace et al. 1987; Tenhunen et al. 1987, 1994; Ryel et al. 1994, 1996; Falge et al. 1997). Simulations were conducted for rainforest canopies with all foliage intact, or with 10 % of the foliage removed in the upper, middle or lower 20 % of the canopy, and with 10 % of the foliage removed uniformly. Total canopy foliage was assumed to be 6 m^2 leaves per m^2 ground area, an approximation based on measurements discussed in Chapter 6. Parameters for single-leaf physiology were measured using gas-exchange for leaves of *Pseudobombax septenatum*, a species with a rather typical C$_3$ physiology (Table 28). Simulations were conducted on a sunny day during early December (end of the

Table 28. Parameter values for leaf-physiology model (Harley et al. 1992; Tenhunen et al. 1990) used in simulations

Parameter	Value	Units	Parameter	Value	Units
Electron transport			**Carboxylase capacity**		
$c (P_{ml})$	45	$\mu mol\ m^{-2}\ s^{-1}$	$c (V_{cmax})$	80	
$\Delta H_a (P_{ml})$	50,000	$J\ mol^{-1}$	$\Delta H_a (V_{cmax})$	50,000	$J\ mol^{-1}$
$\Delta H_d (P_{ml})$	150,000	$J\ mol^{-1}$	$\Delta H_d (V_{cmax})$	150,000	$J\ mol^{-1}$
$\Delta S (P_{ml})$	489	$J\ K^{-1}\ mol^{-1}$	$\Delta S (V_{cmax})$	489	$J\ K^{-1}\ mol^{-1}$
Light utilization			**Carboxylase kinetics**		
α	0.04	$mol\ CO_2$ mol^{-1} photons	$f (K_c)$	299	μbar
Dark respiration	$E_a (K_c)$	65,000	$J\ mol^{-1}$		
$f (R_d)$	1.0	$\mu mol\ m^{-2}\ s^{-1}$	$f (K_o)$	160	mbar
$E_a (R_d)$	50,000	$J\ mol^{-1}$	$E_a (K_o)$	36,000	$J\ mol^{-1}$
Stomatal conductance					
G_{fac}	7.5	-			

Abbreviations:

$c (P_{ml})$	Scaling factor for CO_2- and light-saturated photosynthesis (P_{ml})
$\Delta H_a (P_{ml})$	Activation energy of P_{ml}
$\Delta H_d (P_{ml})$	Deactivation energy of P_{ml}
$\Delta S (P_{ml})$	Entropy factor
α	Initial slope of the light response of CO_2-saturated photosynthesis; light use efficiency
$f (R_d)$	Scaling factor for dark respiration (R_d); (Arrhenius function)
$E_a (R_d)$	Activation energy of R_d
$f (K_c)$	Scaling factor for the Michaelis-Menten constant of carboxylation (K_c)
$E_a (K_c)$	Activation energy of K_c
$f (K_o)$	Scaling factor for the Michaelis-Menten constant of oxygenation (K_o)
$E_a (K_o)$	Activation energy of K_o
$c (V_{cmax})$	Scaling factor for maximum carboxylation rate (V_{cmax})
$\Delta H_a (V_{cmax})$	Activation energy of V_{cmax}
$\Delta H_d (V_{cmax})$	Deactivation energy of V_{cmax}
$\Delta S (V_{cmax})$	Entropy factor
G_{fac}	Sensitivity of stomatal aperture to changes in photosynthesis (see text)

rainy season), although similar results would be expected using any time period.

The results of the simulations strongly suggest that differences in whole canopy primary production were minimal among all foliage removal treatments. Total daily net canopy photosynthesis was calculated as 371.2 mmol m^{-2} day^{-1} for the unforaged canopy. A 10 % foliage removal in the upper 20 % of the canopy changed this value to 358.5 mmol m^{-2} day^{-1}, while removal of the same amount of foliage in the middle or lower canopy led to values of 371.7 and 375.0 mmol m^{-2} day^{-1}, respectively. Uniform removal of 10 % foliage throughout the entire canopy resulted in a canopy photosynthesis of 369.3 mmol m^{-2} day^{-1}. Differences in whole canopy net photosynthesis for canopies with foliage removed as compared to the unforaged canopy were 3 % or less for all simulations. Either slight increases or decreases in whole canopy photosynthesis could result from foliage removal because the canopy was severely light limited. Whether there was an increase or decrease depended on the quantity of foliage operating below the compensation point. Foliage removal reduced the quantity of foliage not exposed to sufficient light for positive net photosynthesis, and thus could result in higher overall canopy photosynthesis. However, if the leaf physiology was such that all foliage had positive net photosynthesis, then foliage removal reduced the whole canopy photosynthesis rate slightly. Understory species would be expected to have a positive carbon gain during the simulated day, and thus have foliage operating above the compensation point. In contrast, the lower leaves of overstory species may operate below the compensation point, and be retained by the tree to shade competitors (Beyschlag et al. 1994). In either case, whole canopy photosynthesis rates differed minimally among harvest treatments.

While foraging by leaf-cutting ants has little effect on whole canopy photosynthesis, the portion of the canopy fixing the most carbon changes with foliage removal. This is most dramatic when foliage is removed from the upper canopy (Fig. 59), the pattern observed for *Atta*. In the unforaged

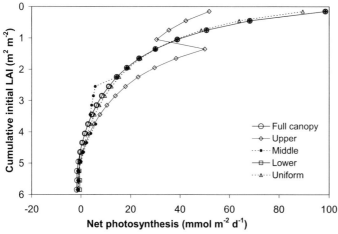

Fig. 59. Simulated daily net photosynthesis rates by canopy depth (expressed as cumulative LAI for full canopy) for a rainforest canopy tree with an initial total leaf-area index of 6-m^2 foliage area per m^2 ground area with different patterns of foliage removal. Simulations consider the removal of 10 % of the total foliage from: the upper 20 % of the canopy (*upper*), the middle 20 % of the canopy (*middle*), the lower 20 % of the canopy (*lower*), and uniformly from the whole canopy (*uniform*). Also shown is daily net photosynthesis with no foliage removed (*full canopy*)

canopy, the most carbon is fixed in the uppermost leaves. However, when foliage is removed by ants from the upper 20 % of the canopy, the portion of the canopy fixing the most carbon extends downward and photosynthesis rates increase below the foraging zone. Removal of foliage in the middle and lower canopy has no effect on the pattern of carbon assimilation above the forage zone and little effect below this zone. Uniform removal of foliage has only a minor effect on the pattern of carbon assimilation, reducing rates slightly in the upper layers, and increasing rates slightly in the lower layers – the former due to reduced foliage and the latter due to increased light availability.

12.2.2 Simulated Effects on Photosynthesis of Single Trees

Within the upper canopy, leaf-cutting ants were estimated to remove up to 40 % of the foliage of a single tree (Chap. 11.4). This foliage loss reduces the photosynthetic potential of the whole tree. However, the loss is not directly proportional to the amount of leaf area removed. The increased light penetration resulting from foliage removal can increase the photosynthetic rate of the remaining leaves. In the simulations conducted, removal of 40 % of the leaves from the upper 25 % of the canopy (10 % of the entire canopy) resulted in only a 27 % reduction in whole tree photosynthesis. Similarly, if a tree was located wholly within the middle 25 % of the canopy, removal of 40 % of the leaves resulted in a 26 % reduction in photosynthesis for the single tree. These results clearly show that the effects of foliage removal by ants on photosynthesis of single trees can be much greater than the minimal effects measured for the whole canopy.

While foliage removal in overstory trees reduces their potential for carbon fixation, trees or vines with foliage below should exhibit increased photosynthesis rates. This was in fact the case. Foliage located below the canopy layers subjected to ant harvest had simulated photosynthesis rates that were 60–100 % higher than the situation where no foliage was harvested. For leaves within the harvested canopy portion, the increase varied from 5–40 %. These increases were for canopies where 10 % of the total canopy foliage was removed in the upper 20 % of the canopy. Thus, plants unaffected by ants, with foliage within or below overstory species subjected to ant foraging, can realize substantial gains in leaf photosynthesis.

12.3 Compensatory Growth

In the above analyses, it was assumed that foliage harvested by ants was not replaced by the plants soon after removal. This assumption was supported by many incidental observations where compensatory growth did not appear to rapidly replace harvested leaves. However, if plants did replace foliage soon after harvest, the predicted effects on canopy light climate and photosynthesis would be reduced since only minimal foliage would be removed at any one time.

To better understand the potential for compensatory growth in response to ant herbivory, a field study was conducted involving six common forest tree species of either pioneer or late successional stages (Table 29). Two twigs of similar size and leaf numbers were selected on six individuals of each species as control and treatment twigs respectively. All leaf blades of the treatment twigs were then removed except for the central vein on half of the twigs during the middle of the rainy season (August, 1997). This removal of leaf material simulated the pattern of removal by leaf-cutting ants. Total leaf area of both treatment and control twigs was monitored over subsequent months.

For all species, there was essentially no regrowth measured during the first 2 months following the artificial harvest (Fig. 60). When leaf regrowth did commence, it took upwards of 6–7 months for the foliage area to approach that of the control twigs, well through much of the dry season. If these species are representative of many of the rainforest species, then compensatory growth following ant harvest appears to be limited. Additionally, these results support the incidental observations of minimal regrowth.

While trees did not rapidly compensate for foliage lost to ant foraging, other plant species may have increased foliage in response to the increased light availability. Increases in foliage area could be expected due to the substantial increases in photosynthesis rates for trees below the zone of ant harvest (cf. Sect. 12.2.2). Minimal differences in average light transmittance to the forest floor between areas with (1.8±1.3 SD, $n=155$) and without (2.2±1.2 SD,

Table 29. Characterization of the tree species used for the defoliation experiment

Species	Family	Successional status	Host plant for *A. colombica*
Annona spraguei	Annonaceae	Pioneer	Yes
Luehea seemannii	Malvaceae	Pioneer	No
Miconia argentea	Melastomataceae	Pioneer	Yes
Anacardium excelsum	Anacardiaceae	Late successional	Yes
Inga goldmanii	Leg.-Mimos.	Late successional	No
Swartzia simplex var. *ochnacea*	Leg.-Caesalp.	Late successional	Yes

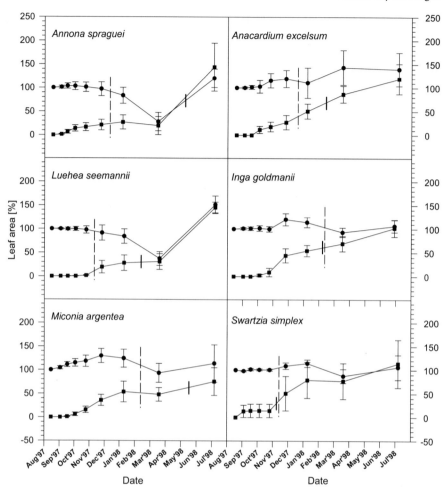

Fig. 60. Leaf area development (mean values ±SE) of controls (*filled circles*) and treated (*filled squares*) twigs of six different neotropical tree species after simulated herbivory (see text for details). *Left of the solid vertical line* the area values of treatment twigs differed significantly from the values before foliage removal, i.e. leaf area was not yet replaced. *Left of the dotted vertical line* values of controls and treatment differed significantly (paired t-tests; $p<0.05$, $n=3–6$)

$n=225$) ant foraging suggest that some middle and lower canopy species may have increased foliage area. While we do not have direct evidence for this, possible candidate species for this increased growth include lianas and young upper canopy species, both of which would likely grow rapidly in response to increased carbon fixation rates resulting from greater incident light. If so, the effect of ant foraging may be to shift the distribution of foliage from preferred harvest species to other plant species. Compensatory growth could then be considered to occur at the community level, not the species level.

The amount of compensatory growth after an herbivory event depends on numerous things. The most important factors are (1) the maximum growth rate of the trees under optimal conditions, (2) the ability to mobilize and retranslocate stored components, (3) the amount and the frequency of the leaf loss (4) the phenological state of the plant at the time of herbivory and (5) resource availability after the herbivory event (Crawley 1983; Oesterheld and McNaughton 1991; Whitham et al. 1991; Trumble et al. 1993). From this list it becomes rather obvious that the compensatory growth behavior of plants after herbivory is strongly influenced by species-specific components. Therefore, the observed differences between species in Fig. 60 are not surprising. However, the observed extended time span between defoliation and full recovery is contradictory to other defoliation studies (e.g., Brown and Ewel 1988). The reason for this is probably the rather low amount of foliage removal compared to the total leaf area of the trees. Nevertheless, the foliage removal pattern in our defoliation study is typical for tropical rainforests where the herbivores are frequently small in size and normally occur at low population densities (Marquis and Braker 1994). Typically monkeys, sloths and the group foraging leaf-cutting ants *Atta* and *Acromyrmex* usually remove far less than 50 % of the total foliage of a tree individual (Rockwood 1976; Milhahn 2001; present study, but see Chap. 11.4) and, the size of the areas damaged by these animals is relatively small with only a few branches in a tree crown are affected (Howard 1990; Weigelt 1996; present study). Large-scale defoliation calamities, such as those caused by herbivorous butterfly larvae, are extremely rare in this ecosystem (e.g., Wong et al. 1990), and total defoliation of entire trees by leaf-cutting ants seems to be restricted to plantations (Cherrett 1986) and has never been reported for natural rainforests.

In general, it appears that compensatory growth by most plants affected by typical leaf-cutting ant herbivory is not immediate and occurs over an extended period of time. While the effects of leaf-cutting ant activity on individual trees are often relatively small, the resulting persistence of canopy gaps can have a pronounced effect on the light microenvironment within and below the canopy. This altered light climate may, therefore, be relevant at the system level in this light limited ecosystem (see Chap. 15)

13 Seed Dispersal by Leaf-Cutting Ants

Ants as dispersal agents represent a well-studied topic mainly in the case of myrmecochorous plants which provoke seed removal by the 'elaiosome', a seed-born appendage serving as protein and oil-rich food reward for the ants (Beattie 1985). This mode of dispersion is common among herbs of temperate mesic forests in the Northern Hemisphere, and woody shrubs in the dry sclerophyll vegetation of Africa and Australia (Beattie 1983). More recently, however, seed dispersal of nonmyrmecochorous plants has become the subject of studies in tropical forests, where ants may comprise up to one third of the entire insect biomass (Fittkau and Klinge 1973), many of them living on the ground (Hölldobler and Wilson 1990) feeding obligatory or occasionally on fruits or seeds (Kaspari 1993, 1996; Levey and Byrne 1993; Horvitz and Schemske 1994; Pizo and Oliveira 1998).

Although it has long been known that leaf-cutting ants occasionally use fruit (or fruit parts) as substrate for their fungus (Plate 16; Cherrett 1968; Weber 1972b), there is only limited information on the extent and the consequences of this behavior and detailed quantitative studies are lacking (Alvarez-Buylla and Martinez-Ramoz 1990; Kaspari 1993; Farji-Brener and Silva 1996; Passos and Ferreira 1996; Leal and Oliveira 1998; Farji-Brener and Medina 2000). Three possible roles of *Atta* and the so-called lower attine ants for the seed biology of tropical forest plants have been discussed so far: (1) the ants may act as secondary dispersal agents of small seeds of certain pioneer species and, by this, affect the local recruitment pattern of these species (Roberts and Heithaus 1986; Kaspari 1993, 1996; Dalling and Wirth 1998); (2) they may facilitate seed germination of vertebrate-dispersed seeds by removing the pulp from the seeds and precluding fungal pathogens (Oliveira et al. 1995; Farji Brener and Silva 1996; Leal and Oliveira 1998); and (3) they may also be disadvantageous, producing a more aggregated seed distribution pattern than other dispersers of the same plant species (Dalling and Wirth 1998), or, if the seeds serve as substrate for the symbiotic fungus, they are seed predators (Nepstad et al. 1990; Moutinho et al. 1993).

As described earlier (Chap. 8, see also Wirth et al. 1997), nonfoliar plant material can represent up to 50 % of the biomass input into an *Atta* colony during the dry season. During 1 year, colony I exploited fruit pulp of *Quararibea asterolepis, Dipteryx panamensis, Anacardium excelsum, Ficus yoponensis* and *F. obtusifolia.* Further, the ants were also observed to carry considerable numbers of seeds from the latter two species as well as from *Cecropia obtusifolia, Brosimum alicstrum* and *Miconia argentea.* Being aware of this, it seems that the importance of foraging fruits and seeds by *Atta* is generally underestimated and needs further investigation. In the following, we present quantitative data on the quantities of fruits foraged from *Miconia argentea* trees by *A. colombica* and the fate of the respective seeds (see also Dalling and Wirth 1998).

13.1 Fruit Harvest from the Pioneer Tree *Miconia argentea* (Melastomataceae)

Miconia argentea is a common pioneer tree in young forest systems and produces round berries with a diameter of 4–8 mm containing numerous tiny seeds from January to June (Croat 1978). *Miconia* fruit was a fairly constant part of the harvest collected by the leaf-cutting ants of both colonies I and II, particularly in April and May (Dalling and Wirth 1998). Berries were either directly cut from the tree, preferably ripe, but sometimes earlier, or collected from the ground. Analysis of the daily *Miconia* harvest observed in a total of ten 24-h counts yielded highly significant relationships between the *Miconia* harvest per minute during peak activity and the total harvest for the day (Fig. 61). Similar to the leaf fragment harvest (see Chap. 8.4) this correlation was subsequently used to estimate further daily inputs from peak activity counts of the colonies.

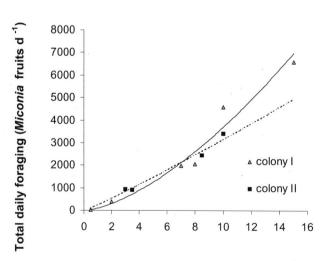

Fig. 61. Relationship between the total number of *Miconia* fruits collected per day and the number of fruits harvested per minute at the peak of daily activity of colonies I ($y=100.77x^{1.5649}$, $r^2=0.99, p<0.001$, df=4; *solid line*) and II ($y=251.3x^{1.1004}$, $r^2=0.97, p<0.05$, df=2; *dotted line*)

Table 30. Number of *Miconia* fruits and corresponding total biomass input per day during the period of fruit harvest of leaf-cutting ant colonies I and II. Fruit collection was observed over 49 days in colony I and 20 days in colony II

Date	Daily input *Miconia* fruits	Total daily input of fragments	*Miconia* input (%)
Colony I			
11.04[a]	1250	273,540	0.46
18.04[a]	2610	204,585	1.28
25.04[b]	1980	204,450	0.97
02.05[b]	4590	187,740	2.44
04.05[a]	8500	–	–
08.05[b]	6590	161,565	4.08
15.05[b]	2070	242,595	0.85
22.05[b]	390	13,365	2.92
29.05[b]	30	11,100	0.27
Daily average (SE)	3110 (1180)		1.7 (1.3)
Estimated total[c]	136,200		
Colony II			
25.04[b]	3420	76,370	4.48
02.05[a]	690	111,760	0.62
04.05[a]	390	–	–
05.05[b]	2460	–	–
08.05[b]	930	119,835	0.78
15.05[b]	900	148,365	0.61
Daily average (SE)	1470 (490)		1.6 (1.9)
Estimated total[c]	28,400		

[a] Daily totals estimated by regression equations given in Fig. 35a.

[b] 24-h count.

[c] Estimated totals over the observation period were obtained by summing up weekly totals derived from daily values.

As shown in Table 30, colony I (48 days) collected almost 140,000 and colony II (20 days) about 29,000 fruits during the course of the harvesting period. Fruit collection activity showed a relatively high variability between observation days and ceased at the end of May. It is remarkable that neither colony invested more than an average of 1.7 % of their total foraging force in collecting fruits.

13.2 The Fate of the Harvested *Miconia* Seeds

Refuse of *Atta* colonies was analyzed for its *Miconia* seed content (see Dalling and Wirth 1998 for methodological details). Between 14 May and 10 June

1994, a mean of 330 *Miconia* germinable seeds/g fresh weight (SE=110) was detected. By 17 July, i.e., about 1 month after the fruiting season, the seed content in the refuse material declined to zero. Seeds collected from the colony refuse revealed a rather high (73 %) germination rate. Nevertheless, this rate was significantly lower than that of seeds directly obtained from fresh fruits which amounted to 84 % (Yates-corrected χ^2=6.38, df=1, p<0.05).

13.3 Ecological Significance of *Atta* Fruit Harvest

The study of Roberts and Heithaus (1986) and the present investigation are so far the only sources for quantitative data of fruit harvest from tropical tree species by leaf-cutting ants. There is, however, another study on the more primitive members of the fungus-growing ants showing that various members of these species seem to play a relevant role in the seed biology of the Brazilian Cerrado by dispersing seeds and enhancing germination of a wide variety of plants of this vegetation (Leal and Oliveira 1998).

The present results reveal that a considerable amount of *Miconia argentea* fruit is transported into the nest and that the seeds remain viable after the fruit pulp is decomposed by the symbiotic fungus. In the particular case of *Atta colombica,* seeds are deposited together with exhausted fungus material on refuse dumps outside the nest. Since other *Atta* species place the refuse in subterranean chambers, where seeds would probably not find suitable conditions for germination or would even rot, this circumstance is crucial to evaluate the role of *A. colombica* as potential dispersal agents. Disposal of seeds in the aboveground refuse pile typically results in a highly aggregated seed distribution pattern of this species (Dalling et al. 1998) and may, therefore, be interpreted to negatively affect the fitness of this pioneer tree as it limits opportunities for seeds to encounter gaps. Nevertheless, there are several lines of evidence to conclude that the overall effect of *Atta colombica* for this tree species is still beneficial. First, after being brought to the refuse dumps, seeds may be postdispersed either by litter ants, including lower attines (Leal and Oliveira 1998), or passively by heavy rains which were observed to spread out the refuse which is frequently piled on slopes next to the nest. Secondly, the foraging efficiency of the ants is limited. Significant numbers of ant loads are lost along the extensive trails from the source tree to the nest (up to 100 m between *Miconia* trees and nest entrances of colony I). This has been analyzed quantitatively for leaf fragments cut by *Atta colombica,* suggesting that 30 % of the fragments did not reach the nest entrance (Lugo et al. 1973). Similar qualitative observations were made in the present study for fruits during the fruiting period of the respective trees. Further, the fact that the fruit pulp is removed from the seeds by the *Atta* fungus points to the possibility of facilitated germination due to reduced probability of fungal infestations. Germination success of seeds left within a *Miconia* fruit was not measured in the present study, but Oliveira et al. (1995) demonstrated that seed cleaning (removal of fruit matter) by attine ants reduces fungal attacks and facilitates seed germination in the mammal-dispersed tree *Hymenaea courbaril*. Likewise, similar results were found by Leal and Oliveira (1998) for attine ants and seeds of seven "Cerrado" species. Thus, the passage through an *Atta* nest may considerably increase the number of seeds escaping from disproportionate mortality near parent plants due to density-dependent predation, pathogen attack, or seedling competition (Janzen 1970; Connell 1971). Finally, abandoned nest sites of *Atta colombica* may represent an important regeneration site for *Miconia*. This hypothesis is supported by a recent study by Farji-Brener and Medina (2000), who found seven times higher densities of germinable *Miconia* seeds around the *Atta* nest than in the adjacent forest soil. Further, abandoned nests typically represent rapidly fluctuating microsites (see Chap. 15) with increased nutrient content (Haines 1975; Farji-Brener and Silva 1995), higher light availability (Farji-Brener and Medina 2000), and reduced competitive status due to the clearing of understory vegetation (Garrettson et al. 1998). All this indicates that they may in fact serve as regeneration islands for this small-seeded gap specialist tree species (Brokaw 1987).

Despite the increasing evidence to ultimately conclude an important role of *Atta colombica* for the regeneration of *Miconia argentea* and other tree species and, hence, for the dynamics of the forest structure and composition, future studies should particularly be concerned about the following questions: (1) how do the activities of other frugivores such as howler monkeys (*Alouatta palliata*) and white-faced monkeys (*Cebus capucinus*), which often throw down whole *Miconia* fruit stands (Hladik and Hladik 1969), affect the foraging patterns of leaf-cutting ants, or to what extent do leaf-cutting ants compete directly with vertebrate frugivores for ripe fruit in the canopy? (2) Does seed cleaning of leaf-cutting ants really diminish seed mortality by fungal pathogens growing on the fruit matter? (3) How large is the proportion of fruits lost along the foraging trails? (4) Does the distribution of current or abandoned *Atta* colonies and the distribution pattern of host trees affect each other? (5) Can abandoned nest sites provide suitable conditions for seedlings of *Miconia* or other harvested species, or is seedling survival generally reduced by the profusion of fine roots of the surrounding trees accessing the accumulation of nutrients as suggested by Haines (1975)?

14 Nutrient Cycling and Water Relations

14.1 Nutrient Cycling

Nutrient content of the vegetation in the humid tropics does not necessarily reflect the nutrient paucity of tropical soils. Acquired nutrients are retained very tightly within the vegetation and reacquisition from decomposition of dead biomass is very effective (for reviews see Vitousek and Sanford 1986; Bruijnzeel 1990; Grubb 1995). In undisturbed primary forests, the amount of nutrients lost through leaching from soil is typically less or equal to amounts added by precipitation. By far the greatest proportion of the annual nutrient requirement of the forest is satisfied by nutrients that are cycled within the forest. Because of the limited availability of free nutrients, nutrient cycling is a key process in tropical rainforests.

14.1.1 Herbivores and Nutrient Cycling

Dead plant material is normally degraded by herbivores and saprophages which enhances the circulation of vital nutrients needed for plant growth. As in most other ecosystems, herbivores consume only a relatively small proportion of the aboveground biomass (i.e., less than 10%; cf. Landsberg and Ohmart 1989). Nevertheless, their importance as regulators of primary production and nutrient cycling has been stressed by various authors (Chew 1974; Mattson and Addy 1975; Owen and Wiegert 1976; Bianchi et al. 1989; Mazancourt et al. 1998). Several mechanisms have been put forward to explain the tight coupling between consumers, nutrient cycling, and the performance of primary producers (for reviews see Bianchi and Jones 1991; Mattson and Addy 1975). Herbivores can (1) accelerate the leaching rate of mobile nutrients from plant foliage, (2) increase the rate of nutrient-rich litter fall, (3) stimulate the redistribution of nutrients within plants from sinks (such as trunks and branches) to source elements like leaves, buds and flowers, with high turnover rates, (4) promote the activity of decomposer organisms, (5) influence the evolution and short-term induction of defensive compounds (Ehrlich and Raven 1964; Haukioja 1990), which in turn inhibit decomposers or their enzymes (Swift et al. 1979), and (6) interrupt withdrawals of nutrients from ageing and dead tissues (i.e., internal nutrient cycling; cf. Perry 1994), thereby increasing turnover rates.

Nutrient cycling by herbivores has been the predominant argument for the so-called grazing optimization hypothesis (McNaughton 1979; Dyer et al. 1986), which assumes a mutualistic interaction between consumers and plants leading to increased resource production and supply. The experimental evidence for this hypothesis has been criticized (Belsky 1986), but several models have been used to show that nutrient cycling by herbivores is a plausible theoretical explanation for grazing optimization of plant primary production (e.g., Dyer et al. 1986; Holland et al. 1992). However, there is a controversy about the conditions under which grazing optimization might occur. According to Loreau's (1995) model, grazing optimization is likely to occur in stable natural systems if herbivores sufficiently increase the ecosystem rate of nutrient turnover and if the total amount of nutrients in the system is sufficiently high. In contrast, Mazancourt et al. (1998) concluded that an increase in nutrient turnover is not sufficient in the long term. Instead, they concluded that consumers should increase primary production in systems which suffer large losses of limiting nutrients during recycling of plant detritus.

Unlike most other insect herbivores, the foraging of *Atta* ants causes a very patchy pattern of nutrient accumulation. These ants concentrate large quantities of freshly cut vegetation from a large forage area into nest chambers with depths to 7 m below the surface (Waller and Moser 1990) where it is degraded by the mutualistic fungus. The remains of the exhausted fungal material, along with dead and dying ants, are deposited either on the soil surface outside the nest (as in the case of *A. colombica*), or in large refuse chambers below the fungus chambers (most other *Atta* species). Because of the enormous nest dimensions (e.g., Stahel and Geijskes 1939; Jonkman 1980b) and the large quantities of harvested plant material (see Chap. 8), leaf-cutting ants have been discussed as an important factor in mineral cycling and organic decomposition processes in tropical ecosystems (Weber 1972b; Alvarado et al. 1981). Nevertheless, concrete evidence from field data is rather sparse. Only Haines has (1975, 1978) provided information on nutrient enrichment through *Atta colombica* in Panama. Nutrient flow per m² refuse dump averaged for 13 elements (S, N, P, K, Ca, Mg, B, Cu, Zn, Sr, Ba, Mn, Na) was about 48 times the flow in the surrounding leaf litter of the forest floor. Most of these nutrients appeared to be recycled to the vegetation as was reflected by a fourfold enhancement of fine root production within the dump compared to the general forest floor (Haines 1978; Farji-Brener and Medina 2000). However, the intensity of recycling may be lower or may even result in a net drain on the ecosystem for leaf-cutting ant species with internal dumps (Haines 1983). With the exception of Haines (1975, 1978) and some experimental studies on host plant selection (Hubbell and Wiemer 1983; Berish 1986; Howard 1987), there are no available data on the mineral composition of fresh and exhausted fungal substrate in nests of leaf-cutting ants.

Further, the effects of such nutrient accumulation on the vegetation are not well understood. Haines (1975) found evidence that the fine root production of trees close to the nest inhibits the growth of seedlings. Considering the high frequency of colony turnover (see Chap. 7), understory plants and tree seedlings may be negatively affected. In contrast, Garrettson et al. (1998) found that abandoned *Atta cephalotes* nests (with subterranean refuse dumps) showed a greater diversity and abundance of small understory plants than the nearby forest.

Because of the limited knowledge about nutrient redistribution and enhancement of nutrient cycling by leaf-cutting ants, the present study attempted to fill some of these gaps. The carbon and nitrogen contents of total leaf harvests from several colonies, individual resource species, and refuse material were analyzed during the present study to provide insights to the quantity of nutrients relocated through ant harvesting activities.

14.1.2 Nutrient Concentrations in Harvested Leaves and Refuse

Carbon and nitrogen contents of harvested leaf fragments were quantified for the combined harvest of several leaf-cutting ant colonies (averages from 1 to 7 samples per colony taken throughout 1 year) as well as for fragments cut from six higher ranked host tree and liana species of colony I (from collections during annual harvesting peaks; see Fig. 51). Analysis of dried leaf samples was performed with a commercially available CHNS-analyzer (Vario EL, Elementar Analysensysteme GmbH, Hanau, Germany).

As summarized in Fig. 62, nitrogen concentration in the overall leaf harvest averaged 2.38±0.28 SD %. This may be due to a high proportion of young leaves (see Chap. 11), which are significantly higher in nitrogen (2.54±0.36 SD %) than mature leaf material (2.21±0.24 SD %; t=2.62; df=21, p=0.015). In individual leaf species, nitrogen content ranged from 1.94±0.19 SD % in *Cecropia insignis* to 2.96±0.42 SD % *in Stigmaphyllon hypargyreum,* with only the latter being significantly higher than the overall mean (ANOVA: F=8.14; df=6, 34; p>0.0001; Tukeys HSD-test: p=0.02). Likewise, carbon contents differed between species (ANOVA: F=34.91; df=5, 23; p<0.0001), ranging from 42.28±1.21 SD % in *Hura crepitans* to 48.23±1.16 SD % in *Apeiba membranacea.* Overall carbon concentration in all nests amounted to 47.7±0.9 SD %. Consequently, C/N ratios varied greatly from 16.1±0.7 in *Hura crepitans* to 25.0±2.9 SD in *Cecropia insignis*. The overall C/N ratio of the entire harvests was 20.52±2.57 SD.

The measured values for nitrogen content largely agreed with a subset of six preferred species of *A. cephalotes* in Costa Rica tested by Hubbel and

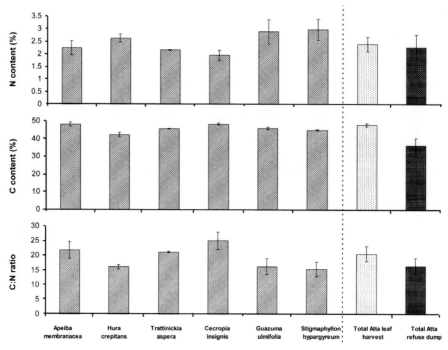

Fig. 62. Nitrogen and carbon content of *Atta colombica* leaf harvest and refuse deposition. Concentrations (as % oven-dry weight) and C/N ratios are given as well as mean values for the entire leaf harvest and refuse output of 12 colonies in the area (averages of 1–7 samples per colony taken throughout 1 year). Data of individual host species are from collections at colony I during annual harvesting peaks as shown in Fig. 25 (number of sampling days: *A. membranacea*: 7; *H. crepitans*: 4; *T. aspera*: 4; *C. insignis*: 6; *G. ulmifolia*: 4; *S. hypargyreum*: 4). *Error bars* = SD. Sampling methods are described in Chapter 8.1

Wiemer (1983), ranging from ca. 1.4 to 2.7%. In the same study, less preferred species showed nitrogen contents between 0.87 and 3.34% N. However, since total nitrogen was not found to correlate with acceptability indices, it appears to be a poor indicator of host plant selection of leaf-cutting ants (Hubbell and Wiemer 1983; Howard 1987; see also Crawley 1983).

Carbon and nitrogen concentrations of the refuse material from the 12 colonies sampled supported this hypothesis. Nitrogen content of exhausted fungus material was only slightly below that of harvested leaf material (5.2%), suggesting that the colony withdraws only minimal nitrogen. In contrast, carbon concentration in the refuse was significantly lower (24.8%) than in leaf input (t=9.38, p<0.0001, df=22). Consequently, passage through the nest significantly decreased the C/N ratio of leaf material from 20.52 to 16.38 (t=3.84, p=0.001, df=22). Thus, by assimilating a considerable proportion of plant carbon rather than nitrogen, leaf-cutting ants and their symbiotic fungus seem to use leaf material primarily as an energy source (Martin and Weber 1969; Bacci et al. 1995). However, reduced C/N ratios in the refuse samples may also at least partially result from high numbers of bacteria and/or other degrading fungi which can frequently be found in the nutrient-rich refuse material. As these destruents usually have C/N ratios between 8 to 10 (Begon et al. 1996), their biomass will of course alter the C/N proportion of fresh refuse.

It is unlikely that total carbon content of leaf material is a good predictor of resource quality since it does not distinguish between the different carbon sources in leaves (cf. Hubbell and Wiemer 1983). As Siqueira et al. (1998) concluded, the metabolic integration involves primarily xylan and starch, whereas cellulose (although it is degraded) seems to be less important for symbiont nutrition. Thus, nonstructural carbohydrates, although largely neglected in host plant selection studies (but see Vasconcelos and Fowler 1990; Nichols-Orians 1991a), seem to be far more relevant than expected.

14.1.3 Nutrient Dynamics Within the Colony

Nutrient contents of leaves and refuse material together with the estimated consumption and refuse deposition rates (see Chap. 8) allowed an estimation of annual fluxes into an average *Atta colombica* nest (Table 31). From a total harvest of approx. 6.5 kg nitrogen and 130 kg carbon, approx. 2 kg of nitrogen and 60 kg of carbon were assimilated by an ant colony and its fungus garden throughout the year and, hence, temporarily immobilized within the system. During the same time more than 4 kg nitrogen and 70 kg carbon accumulated in the external refuse piles.

The proportion of nutrients removed from the overall nutrient pool in the canopy foliage and in the leaf litter was estimated based on published information (although scarce) and the present results (Table 32). Similar to the

Table 31. Mean annual flow of nitrogen (N) and carbon (C) through an *Atta colombica* nest. Estimates are derived from colony consumption rates (see Chap. 8) and mineral concentrations of harvested leaf fragments and refuse material of 12 colonies

	N	C
Total nest input (kg colony^{-1})[a]	6.567	131.25
Total nest output (kg colony^{-1})[b]	4.43	70.35
Consumption (kg colony^{-1})	2.13	60.9
Immobilized proportion of total intake (%)	32.46	46.4

[a] Assuming a leaf consumption of 275 kg year^{-1} colony^{-1} and 2.38 % N and 47.7 % C content in harvested leaf fragments (own results).

[b] Assuming a refuse deposit of 196 kg year^{-1} colony^{-1} and 2.26 % N and 35.89 % C content in the refuse material (own results).

Table 32. Proportions of carbon and nitrogen harvested from the pools present in the standing leaf crop of the canopy and the annual leaf litter. Comparison of calculations at the scale of the colony foraging area and at the 100-ha population scale

	N	C
Total content in standing leaf crop (kg/ha)[a]	123	2690
Content in annual leaf litter (kg/ha)[b]	77	2770
Annual harvest per foraging area of colony I (kg/ha)[c]	12.3	247
Proportion of standing leaf crop (%)	10	9.2
Proportion of annual leaf litter (%)	16.0	8.9
Overall annual harvest of *Atta colombica* population per 100 ha (population scale) (kg/ha)[d]	2.3	46.9
Proportion of standing leaf crop (%)	1.9	1.7
Proportion of total leaf litter (%)	3.0	1.7

[a] Assuming a standing leaf crop of 5697 kg/ha for BCI (estimated from LAI of 5.25, (see Chap. 4) and SLA of 107 g/m^2 from Zotz 1992); leaf N content of 2.2 % [estimated from mature understory leaves from tree saplings (2.35 %; Coley 1983) and own data from mature leaves in leaf-cutting ant harvest (2.21 %)] and a C content of 48 % [from own data from mature leaves in leaf-cutting ant harvest (47.5 %) and Zotz 1992 for three canopy trees (48.5 %)].

[b] According to 6410 kg litter ha^{-1} $year^{-1}$ (Leigh and Windsor 1982), 1.2 % N (Yavitt 2000), and 43.2 % C content (Burchard 1999).

[c] Assuming 517 kg ha^{-1} $year^{-1}$ leaf harvest from discrete foraging area of colony I and mineral contents as given in Table 31.

[d] Assuming 98.4 kg ha^{-1} $year^{-1}$ leaf harvest from 100 ha of forest with an *A. colombica* density of 0.52 ha^{-1} and mineral contents as given in Table 31.

estimation of herbivory rates, the issue of scales is crucial for adequate results. At the 100-ha population level, *Atta colombica* removed <2 % carbon and nitrogen from the total standing leaf crop and ≤3 % of the surrounding leaf litter. On the other hand, at the scale of the foraging area actually used by an individual colony, nitrogen and carbon removal amounted to ≤10 % of the corresponding canopy foliage and 16 and 9 % of the total leaf litter, respectively. As a measure of the amplification of nitrogen and carbon pools by leaf-cutting ant colonies, the concentrations for total nest output given in Table 31 were related to an average nest size of 22 m^2 (see Chap. 7) and compared with the nutrient content within the leaf litter of the forest floor. These analyses indicated that in the colony refuse, nitrogen concentration was 26 times and carbon concentration 12 times higher than in the leaf litter.

14.1.4 Community Level Effects

A possible scenario of the nitrogen and carbon dynamics in the tropical moist forest of BCI as influenced by leaf-cutting ants is provided in Fig. 63. Harvesting 30 % young leaves that have ca. 15 % more nitrogen than mature harvested leaves, leaf-cutting ants select leaves with nutrients that would otherwise have been partly recycled by the tree prior to leaf fall. Although comprehensive data on the aboveground nutrient distribution in the BCI forest are lacking, there is evidence from the present data that the internal translocation of carbon and nitrogen on BCI may be as much as 45 and 10 % of the total foliage nitrogen and carbon pool respectively (see Fig. 63). While it has been shown that species differ greatly in this respect, studies confirm the potential of trop-

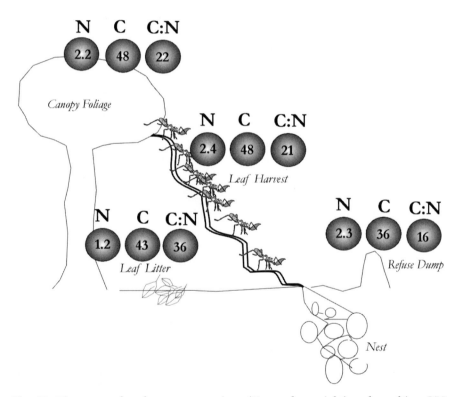

Fig. 63. Nitrogen and carbon concentrations (% per dry weight) and resulting C/N ratios of canopy foliage and leaf litter as affected by *Atta colombica* colonies on BCI. The nutrient content of canopy leaves is a conservative estimate based on own data on the proportion of old leaves in the harvest of ant colonies and published records for N (Coley 1983b for mature leaves of 46 tree saplings: 2.35 %, and L.S. Santiago, (pers. comm.) for canopy leaves of 20 tree species: 1.5–2.5 %) and C (Zotz 1992 for adult canopy leaves of three tree species: 48.5 %). Data for leaf litter were taken from Yavitt (2000) for N, and Burchard (1999) for C; see also Table. 32

ical trees for substantial amounts of internal cycling, even up to 50 % (cf. Bruijnzeel and Proctor 1995; Grubb 1995; L.S. Santiago pers. comm.). Thus, harvesting nutrient-rich young leaves from the trees before the translocation of nutrients into the branches can occur may result in a disproportionate acceleration of nutrient cycling mediated by leaf-cutting ants.

The relatively small proportion of nitrogen immobilized during the fungal degradation process, especially if compared to other herbivores (Crawley 1983), may be due to low nitrogen requirements of the garden fungus (C/N ratio of fungal hyphae on leaves is only ca. 14:1; Swift et al. 1979) or possibly to the efficiency of internal nitrogen recycling by the ants (Martin 1970). In combination with a sizeable proportion of carbon lost through respiration processes or fixation within the ant colony, a considerable reduction of the C/N ratio in the refuse as compared to the surrounding litter (16:1 vs. 27:1) is achieved (Fig. 63). The resulting increase in substrate quality is impressively reflected by the accumulation of a broad diversity of litter arthropods in the refuse dump (Rojas 1989; Burchard 1999) and the high abundance of fine roots that benefit from the increased mineralization rate created by a locally promoted destruent food web (Haines 1978). In consequence, the refuse piles attract many vertebrate predators, such as birds and mammals (Weber 1966) and, hence, serve as a constant resource pool for forest animals.

Generally, the activity of colonies modifies resource flows directly by relocating nutrients and soil particles as well as indirectly through effects on soil properties and feedback mechanisms. In detail, the effects of leaf-cutting ants on the forest nutrient regimes can be summarized as follows:
1. Acceleration of nutrient cycling: plant internal nutrient cycling is reduced by selection of leaves with high nutrient content;
2. Patchy distribution of nutrients on the forest floor: ca. one tenth of the carbon and nitrogen of the canopy within the foraging area of a colony is relocated to the nest area (Chap. 7). Thus, in locations covering less than 0.5 % of the area populated by leaf-cutting ants on BCI, nitrogen flux is about 20–30 times higher than in the rest of the forest.
3. Effects on populations of invertebrates and plant performance: a highly diverse trophic cascade of destruent invertebrates is enhanced through a significant reduction of the C/N ratio in the colony refuse as compared to the surrounding litter. The subsequent mineralization process allows a proliferation of fine roots of nearby plants and, hence, may have positive effects on plant growth and fitness (Haines 1975; Farji-Brener and Medina 2000).

14.2 Water Relations

Although water stress is not generally considered to be a major factor affecting plants of the humid tropics, there is an increasing amount of evidence that the regular dry seasons as well as irregular periods of extreme dryness can lead to pronounced water stress in tropical plants (e.g., Tobin et al. 1999). The intensity and the duration of such dry periods seem to have a considerable effect on growth, mortality, and distribution of the affected species (Condit et al. 1996b; Mulkey and Wright 1996). Under such circumstances, additional herbivory-related water losses may in some cases affect plant water relations.

14.2.1 The Effect of Leaf Wounding

Up to now, potential herbivory-related increases in leaf transpiration have received minimal attention, and such effects have only been demonstrated for soybeans (Hammond and Pedigo 1981; Ostlie and Pedigo 1984). Recently, Herz (2001) has conducted measurements with artificially and similarly wounded leaves of 49 species from different habitats in the neotropical rainforests of Panama and Costa Rica. Using a paper punch, he removed approx. 5 % of the leaf surface. Independent of leaf morphology and anatomy, the minimum leaf conductance (g_{min}) of the wounded leaves was on average 92.5 % higher than the respective values of intact control leaves, with the increase varying from near 0 (four species) to 700 % depending on the species. g_{min} values of the controls varied considerably between the investigated species. Depending on the water saturation deficit of the surrounding air, the water loss resulting from these wounds may be considerable because wound healing was extremely slow. A detailed study on 3 of the 49 investigated species revealed that even 3 weeks after the wounding, treated leaves showed significantly higher g_{min} values than control leaves. Generally, it could be shown that the relative effect of herbivory on leaf water relations increases with the efficiency of the water conserving mechanisms of the leaves characteristic of a particular species.

These results indicate that at the scale of a single twig, herbivory may in fact lower water-use efficiency. Further, recent studies suggest that plants generally operate close to xylem cavitation (Tyree and Sperry 1988; Lösch and Schulze 1994; Sperry et al. 1998). Therefore, an herbivory-related increase in transpiration may under certain circumstances cause the death of the entire branch. This may reduce the competitive ability (through reduced carbon assimilation and shading of competitors) and in consequence the fitness of an individual.

14.2.2 Whole-Canopy Level Effects

In addition to the effects resulting from wounding described above, foliage removal by herbivores should have additional direct and indirect effects on the water relations of plants because (1) transpiring foliage area is removed and (2) previously shaded foliage area in the middle and lower canopy parts becomes exposed to higher light intensities if foliage is removed from the upper canopy areas. The latter will lead to higher leaf temperatures of the remaining foliage, thus increasing the air to leaf vapor pressure difference. Further, higher light levels may also stimulate stomatal opening in the remaining leaves. Finally, wind may penetrate deeper into a canopy which has been thinned out by herbivores. All of these effects will lead to higher canopy transpiration rates. However, if transpiration is limited by the available water (especially during dry periods), the total canopy transpiration may not be greatly affected by foliage loss because water uptake is limited by mass flow to the roots.

Presently, quantitative data about whole-canopy effects on transpiration are lacking. However, since the canopy photosynthesis model used to calculate carbon fluxes (see Chap. 12.2) includes a routine which calculates the concomitant canopy transpiration rates, a rough estimation of some of the effects becomes possible. As was similar for canopy photosynthesis, the simulation

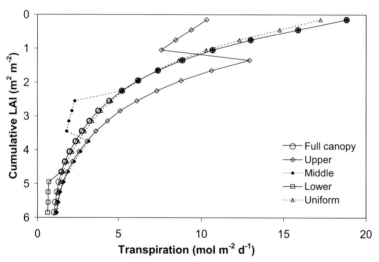

Fig. 64. Simulated daily transpiration rates by canopy depth (expressed as cumulative LAI for full canopy) for a canopy of a *Pseudobombax septenatum* tree with an initial total leaf-area index of 6 m² foliage area per m² ground area with different patterns of foliage removal. Simulations consider the removal of 10 % of the total foliage from: the upper 20 % of the canopy (*upper*), the middle 20 % of the canopy (*middle*), the lower 20 % of the canopy (*lower*), and uniformly from the whole canopy (*uniform*). Also shown is daily transpiration with no foliage removed (*full canopy*)

results strongly suggest that differences in whole-canopy transpiration were minimal among all foliage removal treatments. Total daily canopy transpiration was calculated as 112.2 mol m^{-2} day^{-1} for the unforaged canopy. A 10% foliage removal in the upper 20% of the canopy changed this value to 109.5 mol m^{-2} day^{-1}, while removal of the same amount of foliage in the middle or lower canopy led to values of 109.7 and 110.2 mol m^{-2} day^{-1}, respectively. Uniform removal of 10% foliage across the entire canopy resulted in a canopy transpiration of 109.8 mol m^{-2} day^{-1}. While these results suggest that foraging by leaf-cutting ants has little effect on whole-canopy transpiration, the distribution of the water loss across the canopy changes with foliage removal. This is most dramatic when foliage is removed from the upper canopy (Fig. 64). In the unforaged canopy, most water is transpired in the uppermost leaves. However, when foliage is removed by ants from the upper 20% of the canopy, the portion of the canopy with the highest transpiration rates extends downward and water loss increases below the foraging zone. Removal of foliage in the middle and lower canopy has no effect on the transpiration pattern above the forage zone and a less pronounced, but still clearly visible effect below this zone. Uniform removal of foliage has only a minor effect on the transpiration pattern, reducing rates slightly in the upper layers, and increasing rates slightly in the lower layers – the former due to reduced foliage and the latter due to increased light penetration.

14.2.3 Simulated Effects of Herbivory on Transpiration of Single Trees

Within the canopy of a single tree it is not expected that a reduction of the transpiring foliage due to herbivores will result in a proportional reduction of transpiration. As described above, increased light penetration resulting from foliage removal may actually increase the transpiration rate of the remaining leaves. This contention was supported by model calculations where a 40% foliage removal for a single tree with foliage located within the upper 25% of the canopy (Chaps. 11.4 and 12.2.2) revealed a 28% decrease in whole-tree transpiration. Similarly, calculations for a single tree located wholly within the middle 25% of the canopy with 40% of the foliage removed showed a 31% reduction of transpiration. Nevertheless, these results clearly show that effects of foliage removal by ants on the water relations of single trees can be much greater than the minimal effects calculated for the whole canopy.

The information presented here shows that the activities of herbivores can have significant direct and indirect effects on the water relations of individual plants. Thus, under certain stress conditions, selective herbivores (like leaf-cutting ants) may have a direct effect on the survival and the fitness of individuals. Reduced fitness and survival of individuals of affected species may ultimately have effects on the species composition of the community (cf. Pacala and Crawley 1992).

15 Conclusions: Ecosystem Perspectives

Because of their multifarious effects on the vegetation, leaf-cutting ants have been denoted as key species of Central American rainforest ecosystems (Fowler et al. 1989; Perfecto and Vandermeer 1993). First, they have long been identified as important herbivores in tropical rainforests. As discussed previously (Chap. 12), foliage removal by ants undoubtedly has direct effects on individual plants with some losing up to 40 % of their leaves. However, when scaling this patchily distributed herbivory up to the whole ecosystem (Fig. 52), and comparing it with the estimated 15 % consumption of annual leaf production by all herbivores on BCI (Leigh and Windsor 1982), one is tempted to conclude that the effect of leaf-cutting ants at the ecosystem level is practically negligible. On the other hand, much of the information on the activities of leaf-cutting ants in tropical rainforests, including the material in this book, points in the direction that the overall effects of leaf-cutting ant activity in tropical rainforests may go well beyond the simple removal of foliage. These effects include enhancement of nutrient availability through the enrichment of soil from nest refuse dumps (Haines 1975; Farji Brener and Silva 1995) and the transfer of nutrients to upper soil layers during nest construction (Weber 1972a,b; Perfecto and Vandermeer 1993). Their nests can contain several thousand chambers comprising a total volume of up to 20 m³ (Weber 1966). However, there are also more direct effects of these ants on the vegetation: understory vegetation growing on or overhanging the immediate nest surface is constantly cleared, frequently resulting in understory gaps near nest sites. Moreover, the ants can directly affect vegetation succession through the destruction of numerous flowers of forest tree species (Haines 1975), and can also significantly contribute to seed dispersal of certain forest plants (Roberts and Heithaus 1986; Kaspari 1993, 1996; Dalling and Wirth 1997).

Disturbance can be viewed as having two primary functions in ecosystems, to alter resource availability and to modify the performance of individuals. Activities of leaf-cutting ants can affect both, increasing the spatial and temporal heterogeneity of essential resources (light, nutrients and under certain circumstances even water) and altering the competitive ability of selected individual plants (see below). Both effects are important processes in ecosystem dynamics (White and Jentsch 2001). Further, leaf-cutting ants affect the population of selected species by harvesting reproductive organs and contributing to seed dispersal (Chap. 13). In this final chapter, therefore, we critically evaluate these manifold direct and indirect effects in order to assess the relative importance of leaf-cutting ants to their rainforest ecosystem.

15.1 Disturbance, Herbivory, and Biodiversity

Tropical rainforests are characterized by high species diversity and relatively stable community structure and species composition. This high species richness has been attributed to both disturbance and resource heterogeneity (e.g., Connell 1978; Gentry 1982, 1988; van der Maarel 1993; Givnish 1999; Hubbell et al. 1999; Wright 2002). These two hypothesized mechanisms for species richness are not exclusionary, but in fact complementary. Resource heterogeneity, which is enhanced by disturbance (Platt 1975; Loucks et al. 1985; Canham et al. 1994; White and Jentsch 2001), is linked to niche differentiation and opportunities for more species. Disturbance, in addition to affecting resource heterogeneity, can allow species of different successional characteristics to coexist by returning portions of the ecosystem to earlier successional stages (Vogl 1974; Connell and Slatyer 1977; Noble and Slatyer 1980; Pickett and White 1985).

In the absence of disturbance, ecosystems approach equilibrium and competitive exclusion reduces diversity to minimal levels. When disturbances are very intense or frequent, few species can persist or repeatedly colonize after each disturbance, resulting in low diversity. When disturbances are of intermediate frequency or intensity, there are more opportunities for re-establishment of early to mid-successional species that would otherwise eventually be excluded through competition if the system were allowed to approach equilibrium. Thus, some argue (Connell 1978; Huston 1979) that a peak of diversity should occur at intermediate frequencies and intensities of disturbance, a view portended as the intermediate disturbance hypothesis (Fig. 65). Although a promising perspective, empirical support for this hypothesis is still rather limited. In a meta-analysis of Mackey and Currie (2000), the described "peak" pattern was only found in 19 % of 130 studies reviewed. However, most of these studies contained only qualitative statements rather than hard numbers.

Several authors have suggested that pests increase the diversity of the local environment, effectively increasing the number and diversity of ecological niches for host species in the forest. Janzen (1970) and Connell (1971) developed a model which predicts a tendency for species-specific pests to be more numerous near mature trees. Therefore, the probability of survival of recruits should increase with distance from the tree. This model has been criticized by

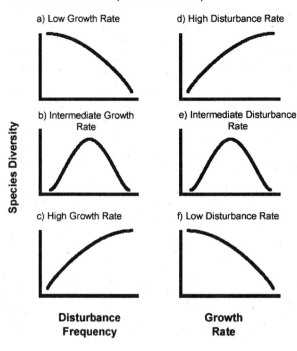

a) Low Growth Rate

b) Intermediate Growth Rate

c) High Growth Rate

d) High Disturbance Rate

e) Intermediate Disturbance Rate

f) Low Disturbance Rate

Species Diversity

Disturbance Frequency

Growth Rate

Fig. 65. The effect of the disturbance regime on species richness depends on the frequency/intensity of disturbance and the rate of population growth/competitive displacement of the community. Intermediate disturbances maximize species richness only at intermediate population growth rates. (Jentsch 2001, after Huston 1994)

Hubbell et al. (1990) and Condit et al. (1992) and, of course, it is not fully applicable to more generalist herbivores. However, leaf-cutting ants as selective herbivores might have this density dependence effect towards the plant species which they harvest. Perhaps this is a mechanism which helps explain the widely dispersed nature of trees of the same species, and leaf-cutting ants could contribute to this. Another density-related model, suggested by Wills (1996), is based on oligophagous rather than monophagous pests. The result is more complex since the performance of a host species depends not only on its own density, but also on the density of other host species that share pest species. Additionally, Wills et al. (1997) demonstrated that density-dependent mortality operates in 80 % of the most common woody plants (>1 cm dbh) in a 50-ha monitoring plot on BCI, a finding which supports the Janzen-Connell hypothesis (Givnish 1999).

The importance of nonequilibrium processes in the maintenance of species diversity has long been emphasized (e.g., Connell 1978; Huston 1979). Nonequilibrium dynamics occur in ecosystems when the return to equilibrium after perturbation is slow in relation to the perturbation frequency. Thus, perturbations prevent complete system equilibrium and the resulting competitive exclusion, and thereby enhance the nonequilibrium coexistence of multiple species (Huston 1979, 1994). The dominance of single plant species is also reduced when similar species become established following a disturbance (Walker et al. 1999). Also, the release of resources following disturbance can shift competitive balances and affect community structure (Brown 1994). Disturbances can result in either more resources available to plants, or less if they are not retained by plants or soils before being lost from the ecosystem (White and Jentsch 2001).

Biotic disturbances which cause mortality in dominant species, thereby enhancing nonequilibrium species coexistence, include predation, herbivory, parasitism, and disease. There is, in fact, evidence that the presence of herbivores can result in higher diversity in algal or plant communities (e.g., Connell 1971; Lubchenco 1978; Huston 1994). Since a generalized predator is likely to feed on the most abundant species in a community, the effect is likely to be an increase in diversity (Huston 1994). The results of the present study (Chap. 10) indicate that leaf-cutting ants may fit this perspective. Generally, insect herbivory has been shown to be one of the disturbance effects which can positively influence secondary plant succession, and thus, species diversity (McBrien et al. 1983; Brown and Gange 1990, 1992; Pacala and Crawley 1992; Davidson 1993; Vasconcelos and Cherrett 1997). In forests, herbivores and pathogens likely increase mortality of shaded seedlings (Augspurger 1984) and subcanopy trees, and also enhance the senescence of early succession dominants. High rates of stand turnover reported from tropical forests (Rankin-de-Morena et al. 1990) may be primarily due to high mortality rates within the smaller trunk diameter classes (Clark and Clark 1992), but such quantitative evidence for rainforest communities is minimal (Denslow 1996).

Although prevalent in ecosystems, resource heterogeneity, like many other ecosystem phenomena, is scale-dependent. An environment which is heterogeneous at one scale may not be heterogeneous at another scale. Thus, the importance of environmental heterogeneity always depends on the scale of response of the organisms of interest (Stewart et al. 2000; Wilson 2000). Simple generalizations such as "heterogeneity enhances biodiversity" are meaningless oversimplifications (Wiens 2000). However, our present understanding is that relatively frequent intermediate disturbances create spatial and temporal heterogeneities in resource availability within ecosystems and can enhance niche separation. Temporal and spatial heterogeneity in resource availability (Hulsmann and Weissing 1999) and niche separation (e.g., Chesson 1985) are likely to enhance coexistence of species. Thus, species which repeatedly cause disturbances within an ecosystem at scales to which plants respond are likely agents helping to maintain higher diversity among plant species. In the following sections we make the case that leaf-cutting ants are among these species.

15.2 Leaf-Cutting Ants and the Spatial and Temporal Heterogeneity of Resources

In many cases, organisms alter the spatial configuration of environments, either by simple occupancy of space (e.g., trees, nests of leaf-cutting ants) or through active modification of the physical or biological structure of the environment. There are many examples of animals both generating and maintaining environmental heterogeneity through activities such as grazing, trampling or affecting soil organisms (Wall and Moore 1999; see also Chaps. 7 and 14). Because these animals forage in a patchy manner, they can alter vegetation structure, growth, production, species composition, and nutrient dynamics at multiple scales (Pickett et al. 2000; Stewart et al. 2000; Wiens 2000).

If one type of organism directly or indirectly modulates the availability of resources to other organisms, this is called "ecological engineering" (Jones et al. 1994; Lawton and Jones 1995). Typical "ecosystem engineers", such as the beaver or prairie dogs have become textbook examples for organisms that can exert far reaching influences on whole communities and the ecological processes acting within them. However, less conspicuous organisms such as harvester ants or termites may also have profound effects, but at finer scales. Virtually all habitats are physically engineered by numerous organisms at different scales, in different ways, with different consequences (Pickett et al. 2000). According to this perspective, leaf-cutting ants would act as "physical engineers" of the rainforest ecosystem, because they directly and indirectly change resource availability for other organisms by forming and changing physical structures within the system. Generally, the "ecosystem engineer" concept seems to be more appropriate for leaf-cutting ants than the "keystone species" concept which was proposed for them by Perfecto and Vandermeer (1993). This is because the latter clearly requires a prominent position of the keystone species within the ecosystem which is obviously not the case for leaf-cutting ants in undisturbed closed forest systems.

Another process by which organisms generate spatial heterogeneity is through functional effects on vegetation boundaries. Boundaries (ecotones, edges) exist as discontinuities between contrasting habitats and have a functional role in the ecosystem because they are a particular type of organismally engineered structure and because organisms interact with them (Pickett et al. 2000). In the case of herbivores, edges can alter herbivore activity such that the effects may be greater or lesser in the forest edge (boundary) than in the intact forest (Cadenasso and Pickett 2000). The present study (Chaps. 7 and 10) provides evidence that edge effects may be involved in both the spatial distribution of the colonies and the harvesting preferences of leaf-cutting ants.

15.2.1 Light Heterogeneity

Removal of leaves by leaf-cutting ants has both spatial and temporal effects on light resources. The mosaic of foraging affects the light climate below the area of ant activity. Foliage is taken in patches and not uniformly. This occurs at the tree level by selection for certain trees and at the within tree level by foraging predominantly in certain parts of the tree canopy (Plate 13). This behaviour increases the size of gaps and creates much more heterogeneity than uniform foraging. Light availability is increased to leaves below the zone of removal, but remains unaffected where ants are not removing foliage. With foliage removal predominately high in the canopy, the potential for changes in light climate related to ant harvesting is maximized. While this probably has little effect on total canopy productivity (Chap. 12), there are shifts in light resources to plants lower in the canopy. Ant foraging also exhibits seasonal variability, both in intensity and plant species selected (Chaps. 8 and 10). These seasonal differences additionally increase temporal heterogeneity in light availability within the foraging areas, and also as compared to unforaged areas. Movement of colonies and subsequent changes in foraging areas adds yet another dimension to temporal heterogeneity in light availability (Chap. 7.2). Increased light below foraged areas can be viewed as an ephemeral window of resource opportunity, which closes when the ants relocate.

At the single leaf level, sun flecks can sometimes provide as much as 80–90 % of total daily PFD, thus contributing significantly to tropical understory light environments (Chazdon et al. 1996). Magnitude, duration, and frequency of sun flecks are largely determined by the physical characteristics of the forest canopy, which is affected by both abiotic and biotic factors (e.g., herbivores). Therefore, understory light environments are highly spatially and temporally variable. Survival and growth of understory plants are certainly strongly dependent on the efficiency with which this heterogeneous resource is utilized for carbon gain. Interspecific variation in this efficiency (light interception, photosynthetic induction and capacity, photoinhibition) can lead to competitive advantages or disadvantages of some species (Watling and Press 2000). Light penetrating through canopy gaps has been linked to higher carbon fixation in understory plants (e.g., Pearcy and Sims 1994), and to expansion opportunities by canopy species both to fill gaps and to reach the overstory (Welden et al 1991; Küppers 1994; Liebermann et al. 1995; Midgley et al. 1995). Although there is abundant information on the existence of these effects, few studies have attempted to quantify the contribution of sun flecks to growth of understory plants (Pearcy 1983; Pfitsch and Pearcy 1992) and how this affects competitive interactions and ultimately species diversity. To date, both positive and neutral effects of sun flecks on plant growth have been documented (Watling and Press 2000).

Disturbance in the form of canopy gaps in forests has been hypothesized to increase species richness in forest canopies (e.g., Ricklefs 1977; Solé and Manrubia 1995; Ryel and Beyschlag 2000). If species were adapted to different light regimes, the light gradients within canopy gaps would show a variety of species with different adaptations. While there is some evidence of differential species responses within gaps (Marks 1974; White et al. 1985), there is little evidence of higher tree species richness within canopy gaps in tropical rainforests (Hubbel et al. 1999). In addition, there is evidence that the tree species composition within rainforest canopy gaps were established prior to gap formation (Welden et al. 1991; Lieberman et al. 1995; Midgley et al. 1995; Hubbel et al. 1999). On the other hand, species variability in light utilization, and the spatial and temporal variation in light availability, should enhance species coexistence. Results for trees of New Zealand old-growth forests (Lusk and Smith 1998) are consistent with this perspective. Using growth ring data from adult trees, they showed that shade-tolerant species went through several episodes of growth release and suppression. This was consistent with exposure to a number of gap events through time. Apparently, these plants were able to survive prolonged periods beneath the canopy and pass through a number of growth release events before reaching the canopy crown. This suggests that variation in dispersal, growth and survival rates of tree species below closed canopies may interact with the stochastic process of gap formation, to effectively shuffle species distribution within a forest (Watling and Press 2000).

The question whether such phenomena can be extrapolated from comparatively large tree-fall gaps where they have predominantly been analyzed to the small canopy gaps created by leaf-cutting ants is difficult to answer. The only available data come from the present study and is in no way sufficient to really make a case. However, the data presented in Chapter 12 showing the relative longevity of ant-related canopy gaps and effects of ant herbivory on the light microclimate at the forest floor indicate at least some similarities. In general, the studies cited above suggest that light heterogeneity within canopy gaps of the size formed by leaf-cutting ants may not directly effect species richness. However, by constantly creating new canopy gaps, they likely contribute to species coexistence, and thus, may help to maintain the existing species diversity.

Foraging activity of leaf-cutting ants is not restricted to leaves in the canopy, and increases in light heterogeneity also result from processes on and near the forest floor (Chap. 12.1). Litter and plants along foraging trails and in the vicinity of nests are actively removed, creating understory gaps and areas of bare soil (Plate 14 above; Farji-Brener and Medina 2000; Howard 2001). In fact, removal of any extraneous material that appears in these cleared areas often begins within minutes of contact by ants (authors' observations on BCI). These open areas are not likely to be of high significance to plants when occupied by ants because of the zealous clearing behavior. However, once ants abandon foraging trails or nests, these open areas provide areas of light penetration to the soil surface (Garrettson et al. 1998) and may affect seed germination (Plates 14 below, 15 above; Vázquez-Yanes and Orozco-Segovia 1994).

15.2.2 Nutrient and Soil Heterogeneity

There is a great amount of variation among plants in the precision of matching resource acquiring organs to the available resources. If the perception and the response of co-occurring species to resource heterogeneity are different, this may affect their interactions and ultimately influence community composition (Fitter et al. 2000; Hutchings et al. 2000; Stewart et al. 2000). A heterogeneous belowground resource distribution can obviously stimulate resource uptake by roots compared to a homogeneous situation as, for example, phosphate uptake by plants is increased when it is concentrated in a small soil fraction rather than uniformly distributed throughout the rooting volume (Jackson and Caldwell 1996), and thus, heterogeneous conditions may support greater plant growth than homogeneous conditions (Hutchings et al. 2000; Stewart et al. 2000; but see Ryel and Caldwell 1998). However, although there are pronounced effects of soil heterogeneity on individual plants, there seem to be only subtle effects at the population level indicating that the plasticity of species may fully compensate for the spatiotemporal heterogeneity in nutrient distribution (Caspar and Cahill 1998; Ryel and Caldwell 1998). Nevertheless, at coarse scales, soil heterogeneity certainly does influence vegetation patterns. For instance, Fitter (1982) showed that vertical soil heterogeneity promoted the coexistence of species, and thus, the diversity of the community. In addition, increases in soil nitrogen have been linked to breaking dormancy and promoting seed germination (Pons 1989). Generally, any heterogeneous resource distribution in the soil will create opportunities for niche differentiation and promote coexistence (Tilman 1982; Fitter et al. 2000; Hutchings et al. 2000; Stewart et al. 2000).

Nutrients in tropical rainforest soils are particularly low in availability to plants, with much of the nutrient supply tied up in the standing biomass (for reviews see Vitousek and Sanford 1986; Bruijnzeel 1990; Grubb 1995). As discussed in detail in Chapter 14.1, harvesting of foliage by ants, however, provides a mechanism by which some of these nutrients can be recycled and concentrated (Haines 1975). Leaf fragments are collected over a large area and concentrated into the fungus producing areas within the nest. Waste from the fungus colony is transferred to refuse piles which have been shown to be relatively rich in nutrients (Plates 10, 11). Leaching and erosion of these refuse piles produces nutrient-enriched soil patches (Haines 1975, 1983). Ant activity within the colony can also result in nutrient enrichment of upper soil layers during nest construction (Weber 1972b) and perhaps after nest abandonment where the majority of the fungus substrate is left intact. Garrettson et al. (1998) found that abandoned nests of A. cephalotes showed a greater diversity (plus marginally greater abundance) of small understory plants relative to the nearby forest. They concluded that ant nests, like canopy gaps, could play an important role in recruitment of new individuals and maintenance of species diversity in neotropical forests (see below Sect. 15.3). Further, increases in the heterogeneity of soil nutrients due to ants are likely to occur in areas near the colony and along main foraging trails where litter and debris are removed. Removal of debris results in reduced material for microbial activity and nutrient recycling in these exposed areas. In addition, there may be a small increase in nutrient concentration adjacent to these cleared areas due to the accumulation of the removed debris. Finally, one could speculate that foliage removal may reduce the nutrient demand of some plants, thereby increasing availability within the region of their roots. As already described for light, temporal heterogeneity is added by seasonally changing harvest preferences and frequent movements of colonies. The enormous amount of soil turnover during the excavation of the new nests (Perfecto and Vandermeer 1993, Herz 2001) which takes place in connection with colony relocation (see Chap. 7.2) is another phenomenon by which leaf-cutting ants increase spatial and temporal nutrient heterogeneity (Plate 15 below).

15.2.3 Heterogeneity of Water Availability

Under certain circumstances, leaf-cutting ant activity may affect water distribution within the rainforest. This can occur both through foliage removal and alteration of the soil surface and upper soil layers. Our simulations indicated that whole canopy transpiration rates may not change much due to leaf removal by ants (Chap. 14.2.2). Transpiration rates of individual trees may or may not be reduced by foliage removal because the relative importance of water leakage through wounded leaves (Chap. 14.2.1) has not effectively been quantified. Nevertheless, it is likely that at least during the dry season, plant water relations may be negatively affected by leaf-cutting ant herbivory (Mulkey and Wright 1996). If trees targeted by ants have lower water use, then water availability within the soil may be affected in the vicinity of roots of these plants. If ants affect the spatial heterogeneity in water availability, then temporal heterogeneity is again added by seasonally changing preferences and colony movements.

15.3 Leaf-Cutting Ants and Ecosystem Biodiversity

Above we have shown that, by being differential herbivores, leaf-cutting ants differentially affect the spatial and temporal resource availability among plants. Disturbance and increased resource heterogeneity caused by leaf-cutting ants at multiple scales likely plays a role in helping to maintain, but not necessarily increase (but see Farji-Brener and Ghermandi 2000), species richness in neotropical rainforests (Fig. 66). This can occur through three mechanisms:

The first mechanism is that the redistribution of resources by ants can alter the competitive interactions among species. Removal of foliage reduces the growth and maintenance potential of preferred trees, and the additional light resources in gaps can be exploited for enhanced carbon fixation by adjacent trees (Welden et al 1991; Küppers 1994; Lieberman et al. 1995; Midgley et al. 1995), epiphytes (Monsi and Murata 1970; Köstner and Lange 1986), and understory species (Oberbauer and Donelly 1986; Chazdon and Pearcy 1991). Loss of leaves reduces a plant's ability to compete for light, and reduced capacity to fix carbon reduces growth, maintenance and reproductive potential (Crawley 1983; Coley et al. 1985; Whitham et al. 1991; Marquis 1992a). In addition to reduced carbon fixation, direct harvesting of flowers may also negatively affect reproductive efforts of plants (Haines 1975). However, there are also studies which reveal an increase in the production of flowers and seeds in response to herbivory (McNaughton 1987; Mathews 1994). Another frequently described effect of herbivory, compensatory growth, may in some cases even overcompensate the herbivore-related foliage loss (Ovaska et al. 1992; Schierenbeck 1994; Honkanen et al. 1994; Oba 1994; Tuomi et al 1994), but this effect seems to be of minimal importance in the forests of the present study (Chap. 12).

In extreme cases, the removal of leaves can result in direct mortality of individual plants. While prevalent in cultivated systems (Cherrett 1986), the mortality of trees from massive foliage removal is extremely rare in relatively

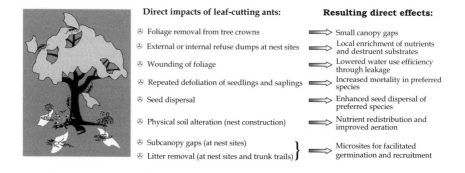

Direct impacts of leaf-cutting ants:	Resulting direct effects:
① Foliage removal from tree crowns	Small canopy gaps
② External or internal refuse dumps at nest sites	Local enrichment of nutrients and destruent substrates
③ Wounding of foliage	Lowered water use efficiency through leakage
④ Repeated defoliation of seedlings and saplings	Increased mortality in preferred species
⑤ Seed dispersal	Enhanced seed dispersal of preferred species
⑥ Physical soil alteration (nest construction)	Nutrient redistribution and improved aeration
⑦ Subcanopy gaps (at nest sites)	Microsites for facilitated germination and recruitment
⑧ Litter removal (at nest sites and trunk trails)	

Resulting indirect effects:

① Reduced fitness of host plants
② Increase of resource heterogeneity
③ Creation of favourable microsites
④ Alteration of competitive balances
⑤ Shifts in population dynamics

Resulting ecosystem level effects:

Affect ecosystem function and help maintain biodiversity of rainforest communities

Fig. 66. Compilation of the various direct and indirect effects of leaf-cutting ants on their ecosystem

pristine native forests (Rockwood 1975). Apparently, defense mechanisms in native plant species limit foraging time, while some cultivated species obviously lack such mechanisms to prevent complete foliage removal. However, among native species, mortality of tree saplings of species preferred by ants can be attributed to foliage removal (Plate 12; Vasconcelos and Cherrett 1997; Rao et al. 2001).

Disadvantaged individuals of tree species preferred by the ants may result in competitive gains by individuals of other species, which in turn may more likely persist. Similarly, ant-related concentrations of nutrients can advantage plant individuals which gain access, and this may allow some species to gain competitively against species which are normally more dominant (Garrettson et al. 1998). One might speculate that tree species preferred by ants put more energy into dominance over tree species which expend energy in defense mechanisms against ants.

A second way leaf-cutting ants can affect species richness is by affecting seedling and sapling survival (Janzen 1970; Connel 1971; Augspurger 1984; Clark and Clark 1992; Vasconcelos and Cherrett 1997; Rao et al. 2001). By reducing the survival rates of preferred species, other species have a greater chance of becoming established or taking advantage of canopy gaps once they open. Movements in ant foraging activity would make the exploitation of such seedlings patchy, creating differences among areas in the species most competitive. Excessive removal of preferred seedlings can have the opposite effect, actually reducing species richness (Rao et al. 2001).

The third way leaf-cutting ants can affect species richness is by affecting seed germination and seedling establishment. Many plant species in rainforests seem to have similar functional roles to other species in the ecosystem. One aspect that seems to differentiate many species with similar function is great variation in reproductive strategy. Differences in seed germination requirements, dispersal, production rates and season of production may allow many seemingly redundant species to become established and coexist simultaneously (e.g., Huston 1994; Hubbell et al. 1999). Heterogeneity in light, nutrients, and soil water induced by ants may help create some of the diversity necessary for different seeds to germinate and for seedlings to become established. In addition, seed dispersal by ants (Chap. 13) may aid in matching seeds to favorable conditions.

Finally, leaf-cutting ant colonies are directly linked to ecosystem diversity via their internal or external refuse dumps which represent impressive hot spots for the diversity of litter arthropods (Rojas 1989; Burchard 1999).

The above discussion makes clear that quantifying individual plant and ecosystem level effects of leaf-cutting ant activity is difficult. Direct mortality of adult trees from ant activity is rare, and issues concerning ecosystem function and attributes such as species richness are still heavily debated. Tropical rainforests are complex ecosystems with many interrelationships between their physical and biological components. Teasing out mechanisms certainly presents a challenge for the research ecologist, but at the same time, increased human pressures on rainforest ecosystems necessitates a better understanding of their structure and function for effective management. By analyzing the role of leaf-cutting ants at multiple scales, the present case study contributes another part to this most fascinating puzzle.

References

Alexandre DY (1982) Étude de l'éclairement du sous-bois d'une forêt dense humide sempervirente (Tai, Côte-d'Ivoire). Acta Oecol Oecol Gen 3:407–447

Alvarado A, Berish CW, Peralta F (1981) Leaf-cutter ant (*Atta cephalotes*) influence on the morphology of Andepts in Costa Rica. J Soil Sci Soc Am 45:790–794

Alvarez-Buylla ER, Martinez-Ramos M (1990) Seed bank versus seed rain in the regeneration of a tropical pioneer tree. Oecologia 84:314–325

Anderson C, Jadin JL (2001) The adaptive benefit of leaf transfer in *Atta colombica*. Insect Soc 48:01–02

Anderson C, Ratnieks FLW (2000) Task partitioning in insect societies: novel situations. Insect Soc 47:198–199

Angeli-Papa J, Eymé J (1985) Les champignons cultivés par les fournis Attinae. Ann Sci Nat Bot Biol Veg 7:103–129

Augspurger CK (1984) Pathogen mortality of tropical tree seedlings: experimental studies of the effects of dispersal distance, seedling density, and light conditions. Oecologia 61:211–217

Autuori M (1941) Contribuciao para o conhecimento da sauva (*Atta* spp. Hymenoptera: Formicidae). I. Evolucao do sauviero (*Atta sexdens rubropilosa*, Forel, 1908). Arch Inst Biol Sao Paulo 12:197–228

Autuori M (1950) Contribui ção para o conhecimento da saúva (*Atta* spp – Hymenoptera-Formicidae). V: Número de formas aladas e redução dos sauveiros iniciais. Arch Inst Biol Sao Paulo 19:325–331

Autuori M (1956) La fondation des sociétés chez les fourmis champignonnistes du genre "*Atta*" (Hymenoptera, Formicidae). In: Autuori M (ed) L'instinct dans le comportement des animaux et de l'homme. Masson et Cie, Paris, pp 77–104

Bacci M, Anversa M, Pagnocca F (1995) Cellulose degradation by *Leucocoprinus gongylophorus*, the fungus cultured by the leaf-cutting ant *Atta sexdens rubropilosa*. Antonie van Leeuwenhoek 67:385–386

Bailey IW (1920) Some relations between ants and fungi. Ecology 1:174–189

Ball MC, Cowan IR, Farquhar GD (1988) Maintenance of leaf temperature and the optimisation of carbon gain in relation to water loss in a tropical mangrove forest. Aust J Plant Physiol 15:263–276

Barone JA (1994) Herbivores and herbivory in the canopy and understory on Barro Colorado Island, Panama. Selby Bot Gard 1st Int Canopy Conf Sarasota, FL (Abstract)

Bass M, Cherrett JM (1994) The role of leaf-cutting ant workers (Hymenoptera: Formicidae) in fungus garden maintenance. Ecol Entomol 19:215–220

Bass M, Cherrett JM (1995) Fungal hyphae as a source of nutrients for the leaf-cutting ant *Atta sexdens*. Physiol Entomol 20:1–6

Bass M, Cherrett JM (1996) Leaf-cutting ants (Formicidae, Attini) prune their fungus to increase and direct its productivity. Funct Ecol 10:55–61

Basset Y, Aberlenc HP, Delvare G (1992) Abundance and stratification of foliage arthropods in a lowland rain forest of Cameroon. Ecol Entomol 17:310–318

Bazire-Benazet M (1957) Sur la formation de l'oeuf alimentaire chez *Atta sexdens rubropilosa*, Forel, 1908 (Hymenoptera, Formicidae). CR Acad Sci Paris 244:1277–1280

Beattie AJ (1983) Distribution of ant-dispersed plants. Sonderbd Naturwiss Ver Hamburg 7:249–270

Beattie AJ (1985) The evolutionary ecology of ant-plant mutualisms. Cambridge University Press, Cambridge

Begon M, Harper JL, Townsend CR (1996) Ecology: individuals, populations and communities. Blackwell, Boston

Berish CW (1986) Leaf-cutting ants (*Atta cephalotes*) select nitrogen rich forage. Am Midl Nat 115:268–276

Belsky AJ (1986) Does herbivory benefit plants? A view of the evidence. Am Nat 127:870–892

Beyschlag W, Ryel RJ, Dietsch C (1994) Shedding of older needle age classes does not necessarily reduce photosynthetic primary production of Norway spruce. Analysis with a 3-dimensional canopy photosynthesis model. Trees Struct Funct 9:51–59

Bianchi TS, Jones CG, Shachak M (1989) The positive feedback of consumer population density on resource supply. Trends Ecol Evol 4:234–238

Bianchi TS, Jones CG (1991) Density-dependent positive feedbacks between consumers and their resources. In: Cole J, Lovett G, Findlay S (eds) Comparative analyses of ecosystems. Patterns, mechanisms and theories. Papers from the third Cary Conference held in Millbrook, NY in 1989. Springer, Berlin Heidelberg New York, pp 331–340

Bitancourt AA (1941) Expressão matematica de crescimento de formigueiros de *Atta sexdens rubropilosa* representado pelo aumento do numero de olheiros. Arch Inst Biol São Paulo 12:229–236

Björkman O, Ludlow MM (1972) Characterisation of the light climate on the floor of a Queensland rainforest. Carnegie Inst Wash Yearbook 71:85–94

Björkman O, Ludlow MM, Morow PS (1972) Photosynthetic performance of rain forest species in their native habitat and analysis of their gas exchange. Carnegie Inst Wash Yearbook 71:94–102

Blanton CM, Ewel JJ (1985) Leaf-cutting ant herbivory in successional and agricultural tropical ecosystems. Ecology 66:861–869

Bolton B (1994) Identification guide to the ant genera of the world. Harvard University Press, Cambridge, MA

Bolton B (1995) A new general catalogue of the ants of the world. Harvard University Press, Cambridge, MA

Bongers F, Vander Meer PJ, Oldemann RAA, Schalk B, Sterck FJ (1990) Levels of herbivory in a tropical rainforest canopy in French Guyana. In: Hallé F, Blanc P (eds) Biologie d'une canopée de foret équatoriale. PRDC, Paris, France, pp 166–177

Boomsma JJ, Ratnieks FLW (1996) Paternity in eusocial hymenoptera. Philos Trans R Soc B 351:947–975

Bot ANM, Currie CR, Hart AG, Boomsma JJ (2001a) Waste management in leaf-cutting ants. Ethol Ecol Evol 13:225–237

Bot ANM, Rehner SA, Boomsma JJ (2001b) Partial incompatibility between ants and symbiotic fungi in two sympatric species of *Acromyrmex* leaf-cutting ants. Evolution 55:1980–1991

Bowers MA, Porter SD (1981) Effect of foraging distance on water content of substrates harvested by *Atta colombica* (Guerin). Ecology 62:273–275

Boyd ND, Martin MM (1975) Faecal proteinases of the fungus-growing ant, *Atta texana*: their fungal origin and ecological significance. J Insect Physiol 21:1815–1820

Bradshaw JWS, Howse PE, Baker R (1986) A novel autostimulatory pheromone regulating transport of leaves in *Atta cephalotes*. Anim Behav 34:234–240

Braganca MAL, Tonhasca AJ, Della Lucia TMC (1998) Reduction in the foraging activity of the leaf-cutting ant *Atta sexdens* caused by the phorid *Neodohmiphora* sp. Entomol Exp Appl 89:305–311

Breda JM van, Stradling DJ (1994) Mechanisms affecting load size determination in *Atta cephalotes* L. (Hymenoptera, Formicidae). Insect Soc 41:423–434

Brokaw NVL (1987) Gap-phase regeneration of three pioneer tree species in a tropical forest. J Ecol 75:9–19

Brown BJ, Ewel JJ (1988) Responses to defoliation of species-rich and monospecific tropical plant communities. Oecologia 75:12–19

Brown DG (1994) Beetle folivory increases resource availability and alters plant invasion in monocultures of Goldenrod. Ecology 75:1673–1683

Brown VK, Gange AC (1990) Insect herbivory and its effects on plant succession. In: Burdon JJ, Leather SR (eds) Pests, pathogens and plant communities. Blackwell, Oxford, pp 275–288

Brown VK, Gange AC (1992) Secondary plant succession: how is it modified by herbivory? Vegetatio 101:3–13

Bruijnzeel LA (1990) Hydrology of moist tropical forests and effects of conversion: a state of knowledge review. Free University, Amsterdam

Bruijnzeel LA, Proctor J (1995) Hydrology and biogeochemistry of tropical montane cloud forests: what do we really know? In: Hamilton LS, Juvik JO, Scatena FN (eds) Tropical montane cloud forests. Ecological studies, vol 110. Springer, Berlin Heidelberg New York, pp 38–78

Bryant JB, Chapin III FS, Klein DR (1983) Carbon/nutrient balance of boreal plants in relation of vertebrate herbivory. Oikos 40:357–368

Burchard K (1999) Zusammensetzung und Strukturierung der Arthropodenzönose in den überirdischen Abfallhaufen von *Atta colombica* in Panama. Diploma Thesis, University of Würzburg, Würzburg, Germany

Burd M (1995) Variable load size-ant size matching in leaf-cutting ant, *Atta colombica*. J Insect Behav 8:715–722

Burd M (1996a) Foraging performance by *Atta colombica*, a leaf-cutting ant. Am Nat 148:597–612

Burd M (1996b) Server system and queuing models of leaf harvesting by leaf-cutting ants. Am Nat 148:613–629

Burd M (2000) Body size effects on locomotion and load carriage in the highly polymorphic leaf-cutting ants *Atta colombica* and *Atta cephalotes*. Behav Ecol 11:125–131

Cadenasso ML, Pickett STA (2000) Linking forest edge structure to edge function: mediation of herbivore damage. J Ecol 88:31–44

Caldwell MM, Meister HP, Tenhunen JD, Lange OL (1986) Canopy structure, light microclimate and leaf gas exchange of *Quercus coccifera* L. in a Portuguese macchia: measurements in different canopy layers and simulations with a canopy model. Trees Struct Funct 1:25–41

Campbell GS (1986) Extinction coefficients for radiation in plant canopies calculated using an ellipsoidal inclination angle distribution. Agric For Meteorol 36:317–321

Canham CD, Finzi AC, Pacala SW, Burbank DH (1994) Causes and consequences of resource heterogeneity in forests: interspecific variation in light transmission by canopy trees. Can For 24:337–349

Carroll CR, Janzen DH (1973) Ecology of foraging by ants. Ann Rev Ecol Syst 4:231–257

Caspar BB, Cahill JF (1998) Population-level responses to nutrient heterogeneity and density by *Abutilon theophrasti* (Malvaceae): an experimental neighborhood approach. Am J Bot 85:1680–1687

Chapela IH, Rehner SA, Schlutz TR, Mueller UG (1994) Evolutionary history of the symbiosis between fungus-growing ants and their fungi. Science 266:1691–1694

Chapin III FS, Bloom AJ, Field CB, Waring RH (1987) Plant responses to multiple environmental factors. BioScience 37:49–57

Chason JW, Baldocchi DD, Huston MA (1991) A comparison of direct and indirect methods for estimating forest canopy leaf area. Agric For Meteorol 57:107–128

Chazdon RL, Field CB (1984) Light environments of tropical forests. In: Medina E, Mooney HA, Vasquez-Yanez C (eds) Physiological ecology of plants of the wet tropics. Junk, The Hague, pp 27–36

Chazdon RL, Pearcy RW (1986) Photosynthetic responses to light variation in rainforest species. II. Carbon gain and photosynthetic efficiency during lightflecks. Oecologia 69:524–531

Chazdon RL, Pearcy RW (1991) The importance of sunflecks for forest understory plants. Bioscience 41:760–766

Chazdon RL, Pearcy RW, Lee DW, Fetcher N (1996) Photosynthetic responses of tropical forest plants to contrasting light environments. In: Mulkey SS, Chazdon RL, Smith AP (eds) Tropical forest plant ecophysiology. Chapman and Hall, New York, pp 5–55

Cherix D, Bourne JD (1980) A field study on a supercolony of the red wood ant *Formica lugubris* Zett. in relation to other predatory arthropods (spiders, harvestmen and ants). Rev Suisse Zool 87:955–973

Cherrett JM (1968) The foraging behavior of *Atta cephalotes* L. (Hymenoptera, Formicidae) I. Foraging pattern and plant species attacked in tropical rain forest. J Anim Ecol 37:387–403

Cherrett JM (1972a) Some factors involved in the selection of vegetable substrate by *Atta cephalotes* (L) (Hymenoptera:Formicidae) in tropical rain forest. J Anim Ecol 41:647–660

Cherrett JM (1972b) Chemical aspects of plant attack by leaf-cutting ants. In: Harborne JB (ed) Phytochemical ecology. Academic Press, London, pp 13–24

Cherrett JM (1983) Resource conservation by the leaf-cutting ant *Atta cephalotes* in tropical rain forest. In: Sutton SL, Whitmore TC, Chadwich AC (eds) Tropical rain forest. Ecology and management, vol 2. Blackwell, Oxford, pp 253–263

Cherrett JM (1986) History of the leaf-cutting ant problem. In: Lofgren CS, Vander Meer RK (eds) Fire ants and leaf-cutting ants. Westview Press, Boulder, pp 10–17

Cherrett JM (1989) Leaf-cutting ants. In: Lieth H, Werger MJA (eds) Ecosystems of the world 14B. Elsevier, Amsterdam, pp 473–486

Cherrett JM, Cherrett FJ (1975) A bibliography of the leaf-cutting ants, *Atta* spp. and *Acromyrmex* spp. up to 1975. Overseas Dev Nat Res Inst Bull 14:1–58

Cherrett JM, Pergrine DJ (1976) A review of the status of leaf-cutting ants and their control. Ann Appl Biol 84:124–133

Chesson PL (1985) Coexistence of competitors in spatially and temporally varying environments: A look at the combined effects of different sorts of variability. Theor Pop Biol 28:263–287

Chew RM (1974) Consumers as regulators of ecosystems: an alternative to energetics. Ohio J Sci 74:359–370

Clark DA, Clark DB (1992) Life history diversity of canopy and emergent trees in a neotropical rain forest. Ecol Monogr 62:315–344

Claver S (1990) Methods for estimating the population density of leaf-cutting ant colonies. In: Vander Meer RK, Jaffe K, Cedeno A (eds) Applied myrmecology – a world perspective. Westview Press, Boulder, CO, pp 220–227

Cole BJ (1983) Multiple mating and the evolution of social behaviour in the Hymenoptera. Behav Ecol Sociobiol 12:191–201

Coley PD (1980) Effects of leaf age and plant life history patterns on herbivory. Nature 284:545–546

Coley PD (1982) Rates of herbivory on different tropical trees. In: Leigh EG Jr, Rand AS, Windsor DM (eds) The ecology of a tropical forest seasonal rhythms and long term changes. Smithsonian Institute Press, Washington, DC, pp 123–133

Coley PD (1983a) Herbivory and defensive characteristics of tree species in a lowland tropical forest. Ecol Monogr 53:209–233

Coley PD (1983b) Intraspecific variation in herbivory on two tropical tree species. Ecology 64:426–433

Coley PD (1988) Effects of plant growth rate and leaf life time on the amount and type of anti-herbivore defence. Oecologia 74:531–536

Coley PD, Aide TM (1991) Comparison of herbivory and plant defenses in temperate and tropical broad-leaved forests. In: Price PW, Lewinsohn TM, Fernandes GW, Benson WW (eds) Plant-animal interactions: evolutionary ecology in tropical and temperate regions. Wiley, New York, pp 25–49

Coley PD, Barone JA (1996) Herbivory and plant defenses in tropical forests. Ann Rev Ecol Syst 27:305–335

Coley PD, Bryant JP, Chapin III FS (1985) Resource availability and plant anti-herbivore defense. Science 230:895–899

Condit R, Hubbell SP, Foster RB (1992) Stability and change of a neotropical forest over a decade. BioScience 42:822–828

Condit R, Hubbell SP, Foster RB (1996a) Assessing the response of plant functional types in tropical forests to climatic change. J Veg Sci 7:405–416

Condit R, Hubbell SP, Foster RB (1996b) Changes in tree species abundance in a neotropical forest: impact of climate change. J Trop Ecol 12:231–256

Connell JH (1971) On the role of natural enemies in preventing competitive exclusion in some marine animals and rain forest trees. In: de Boer PJ, Gradwell GR (eds) Dynamicy of populations. Proceedings of the Advanced Study Institute on Dynamics of Numbers in Populations, Oosterbeek, 1970, Netherlands. Centre for Agricultural Publishing and Documentation, Wageningen, pp 298–312

Connell JH (1978) Diversity in tropical rain forests and coral reefs. Science 199:1302–1310

Connell JH, Slatyer RO (1977) Mechanisms of succession in natural communities and their role in community stability and organisation. Nature 111:1119–1144

Connell JH, Lowman MD, Noble IR (1997) Subcanopy caps in temperate and tropical forests. Aust J Ecol 22:163–168

Corso CR, Serzedello A (1981) A study of multiple mating habit in *Atta laevigata* based on the DNA content. Comp Biochem Physiol 69:901–902

Craven SE, Dix MW, Michaels GE (1970) Attine fungus gardens contain yeasts. Science 169:184–186

Crawley MJ (1983) Herbivory: the dynamics of animal plant interactions. Blackwell Science, Oxford

Crawley MJ (1997) Plant ecology. Blackwell Science, Oxford

Cressie NAC (1993) Statistics for spatial data. Wiley & Sons, New York

Croat TB (1978) Flora of Barro Colorado Island. Stanford University Press, Stanford, CA

Cross JH, Byler RC, Ravid U, Silverstein RM, Robinson SW, Baker JS, De Oliveira JS, Jutsum AR, Cherrett JM (1979) The major component of the trail pheromone of the leaf-cutting ant, *Atta sexdens rubropilosa* Forel:3-ethyl-2,5-dimethylpyrazine. J Chem Ecol 5:187–203

Crozier RH, Page RE (1985) On being the right size: male contributions and multiple mating in social Hymenoptera. Behav Ecol Sociobiol 18:105–115

Currie CR (2001a) Prevalence and impact of a virulent parasite on a tripartite mutualism. Oecologia 128:99–106

Currie CR (2001b) A community of ants, fungi, and bacteria: a multilateral approach to studying symbiosis. Annu Rev Microbiol 55:357–380

Currie CR, Stuart AE (2001) Weeding and grooming of pathogens in agriculture by ants. Proc R Soc Lond B 268:1033–1039

Currie CR, Mueller UG, Malloch D (1999a) The agricultural pathology of ant fungus gardens. Proc Natl Acad Sci USA 96:7998–8002

Currie CR, Mueller UG, Malloch D (1999b) Fungus-growing ants use antibiotic-producing bacteria to control garden parasites. Nature 398:701–704

Dalling JW, Wirth R (1998) Dispersal of *Miconia argentea* seeds by the leaf-cutting ant *Atta colombica* (L.). J Trop Ecol 14:705–710

Dalling JW, Hubbell SP, Silvera K (1998) Seed dispersal, seedling establishment and gap partitioning among tropical pioneer trees. J Ecol 86:674–689

Davidson DW (1993) The effects of herbivory and granivory on terrestrial plant succession. Oikos 68:23–35

De la Cruz M, Dirzo R (1987) A survey of the standing levels of herbivory in seedlings from a Mexican rainforest. Biotropica 19:98–106

Denslow JS (1987) Tropical rainforest gaps and tree species diversity. Annu Rev Ecol Syst 18:431–451

Denslow JS (1996) Functional group diversity and responses to disturbance. In: Orians GH, Dirzo R, Cushman JH (eds) Biodiversity and ecosystem processes in tropical forests. Ecological studies, vol 122. Springer, Berlin Heidelberg New York, pp 127–151

Denslow JS, Hartshorn GS (1994) Tree-fall gap environment and forest dynamic process. In: McDade LA, Bawa KS, Hespenheide HA, Hartshorn GS (eds) La Selva-ecology and natural history of a neotropical rain forest. The University of Chicago Press, Chicago, pp 120–127

Dietrich WE, Windsor DM, Dunne T (1982) Geology, climate and hydrology of Barro Colorado Island. In: Leigh EG Jr, Rand AS, Windsor DM (eds) The ecology of a tropical forest: seasonal rhythms and long term changes. Smithsonian Institute Press, Washington, DC, pp 21–46

Diniz LM, Brandão CRF, Yamamoto CI (1998) Biology of *Blepharidatta* ants, the sister group of the *Attini*: a possible origin of fungus-ant symbiosis. Naturwissenschaften 85:270–274

Dirzo R (1984a) Herbivory, a phytocentric review. In: Dirzo R, Sarukhan J (eds) Perspectives in plant population biology. Sinauer, Sunderland, MA, pp 141–165

Dirzo R (1984b) Insect-plant interactions: some ecophysiological consequences of herbivory. In: Medina E, Mooney HA, Vasquez-Yanes C (eds) Physiological ecology of plants of the wet tropics. Junk Publishers, The Hague, pp 209–224

Dixon AFG (1966) The effect of population density and nutritive status of the host plant on the summer reproductive activity of the sycamore aphid, *Drepanosiphum plantanoides* (Schr.). J Anim Ecol 35:105–112

Dyer MI, Turner CL, Seastedt TR (1986) Herbivory and its consequences. Ecol Appl 3:10–16

Edwards PJ, Grubb PJ (1977) Studies of mineral cycling in a montane rain forest in New Guinea I. The distribution of organic matter in the vegetation and soil. J Ecol 65:943–969

Ehrlich PR, Raven PH (1964) Butterflies and plants: a study in coevolution. Evolution 18:586–608

Eibl-Eibesfeldt J, Eibesfeldt E (1967) Das Parasitenabwehren der Minima-Arbeiterinnen der Blattschneider-Ameise (*Atta cephalotes*). Z Tierpsychol 24:278–281

Emery C (1899) Végétarianisme chez les fourmis. Arch Sci Phys Nat 8:488–490

Emmel TC (1967) Ecology and activity of leaf-cutting ants (*Atta* sp.). In: Advanced zoology: insect ecology in the tropics, winter 1967. Organization for Tropical Studies, San Jose, Costa Rica, pp 125–131

Erwin TL (1982) Tropical forests: their richness in Coleoptera and other species. Coleopterist's Bull 36:74–75

Erwin TL, Scott JC (1980) Seasonal and size patterns, trophic structure, and richness of Coleoptera in the tropical arboreal ecosystem: the fauna of the tree *Luehea seemanii* Triana and Planch in the Canal Zone of Panama. Coleopt Bull 34:305–322

Evershed RP, Morgan ED (1982) The amounts of trail pheromone substances in the venom of workers of four species of attine ants. Insect Biochem 13:469–474

Falge E, Ryel RJ, Alsheimer M, Tenhunen JD (1997) Sensitivity of stand structure and physiology on forest gas exchange: a simulation study for Norway spruce. Trees Struct Funct 11:436–448

Farji-Brener AG (1993) Influence of seasonality on the foraging rhythms of *Atta laevigata* [Hymenoptera: Formicidae] in a tropical savanna. Rev Biol Trop 41:897–899

Farji-Brener AG (2001) Why are leaf-cutting ants more common in early secondary forests than in old-growth tropical forests? An evaluation of the palatable forage hypothesis. Oikos 92:169–177

Farji-Brener AG, Ghermandi L (2000) Influence of nests of leaf-cutting ants on plant species diversity in road verges of northern Patagonia. J Veg Sci 11:453–460

Farji-Brener AG, Illes AE (2000) Do leaf-cutting ant nests make "bottom-up" gaps in neotropical rain forests? A critical review of the evidence. Ecol Lett 3:219–227

Farji-Brener AG, Medina C (2000) The importance of where to dump the refuse: seed banks and fine roots in the nest of the leaf-cutting ants *Atta cephalotes* and *Atta colombica*. Biotropica 32:120–126

Farji-Brener AG, Sierra C (1988) The role of trunk trails in the scouting activity of the leaf-cutting ant *Atta cephalotes*. Ecoscience 5:271–274

Farji-Brener, AG, Sierra C (1993) Distribution of attacked plants along trails in leaf-cutting ants (Hymenoptera: Formicidae): consequences in territorial strategies. Rev Biol Trop 41:891–896

Farji Brener AG, Silva JF (1995) Leaf-cutting ants and forest groves in a tropical parkland savanna of Venezuela: facilitated succession? J Trop Ecol 11:651–669

Farji Brener AG, Silva JF (1996) Leaf-cutter ants' (*Atta laevigata*) aid to the establishment success of *Tapiria velutinifolia* (Anacardiaceae) seedlings in a parkland savanna. J Trop Ecol 12:163–168

Farqhuar GD, von Caemmerer S (1982) Modelling of photosynthetic response to environmental conditions. In: Lange OL, Nobel PS, Osmond CB, Ziegler H (eds) Encyclopedia of plant physiology, new series, vol 12B. Springer, Berlin Heidelberg New York, pp 549–587

Farquhar GD, von Caemmerer S, Berry JA (1980) A biochemical model of photosynthetic CO_2 assimilation in leaves of C_3 species. Planta 149:78–90

Febvay G, Kermarrec A (1983) Enzymes digestives de la fourmi attine *Acromyyrmex octospinosus* (Reich): caractérisation des amylases, maltase et tréhalase des glandes labiales et de l'intestin moyen. CR Acad Sci Paris 296:453–456

Feener DH, Moss KAG (1990) Defense against parasite by hitchhikers in the leaf-cutting ants: a quantitative assessment. Behav Ecol Sociobiol 26:17–26

Feener HG, Lighton JRB, Bartholemew GA (1988) Curvilinea allometry, energetics and foraging ecology: a comparison of leaf-cutting ants and army ants. Funct Ecol 2:509–520

Feeny PP (1976) Plant apparency and chemical defense. Rec Adv Phytochem 10:1–40

Filip VDR, Maass JM, Sarukhan J (1995) Within-and among-year variation in the levels of herbivory on the foliage of trees from a Mexican tropical deciduous forest. Biotropica 27:78–86

Fisher PJ, Stradling DJ, Sutton B, Petrini CLE (1996) Microfungi in the fungus gardens of the leafcutting ant *Atta cephalotes*: a preliminary study. Mycol Res 100:544–546

Fisher RA, Corbet AS, Williams CB (1943) The relation between the number of species and the number of individuals in a random sample of an animal population. J Anim Ecol 12:42–57

Fitter A (1982) Influence of soil heterogeneity on the co-existence of grassland species. J Ecol 70:139–148

Fitter A, Hodge A, Robinson D (2000) Plant response to patchy soils. In: Hutchings MJ, John EA, Stewart AJA (eds) The ecological consequences of environmental heterogeneity. The 40th Symposium of the British Ecological Society held at the University of Sussex, 23–25 March 1999. Blackwell Science, Oxford, pp 71–90

Fittkau EJ, Klinge H (1973) On biomass and trophic structure of the central Amazonian rainforest ecosystem. Biotropica 5:2–14

Fjerdingstad EJ, Boomsma JJ (1997) Variation in size and sperm content of sexuals in the leafcutter ant *Atta colombica*. Insect Soc 44:209–218

Fjerdingstad EJ, Boomsma JJ (1998) Multiple mating increases the sperm stores of *Atta colombica* leafcutter ants. Behavioral Ecoloy and Sociobiology 42:257–261

Fjerdingstad EJ, Boomsma JJ, Thorén P (1998) Multiple paternity in the leafcutter ant *Atta colombica* – a microsatellite DNA study. Heredity 80:118–126

Folgarait PJ, Dyer LA, Marquis RL, Braker HE (1996) Leaf-cutting ant preferences for five native tropical plantation tree species and different light conditions. Entomol Exp Appl 80:521–530

Foster RB (1982) The seasonal rhythm of fruitfall on Barro Colorado Island. In: Leigh EG, Rand AS, Windsor DM (eds) The ecology of a tropical forest: seasonal rhythms and long-term changes. Smithsonian Institution Press, Washington, DC, pp 67–81

Foster RB, Brokaw NVL (1982) Structure and history of vegetation of Barro Colorado Island. In: Leigh EG Jr, Rand AS, Windsor DM (eds) The ecology of a tropical forest: seasonal rhythms and long term changes. Smithsonian Institute Press, Washington, DC, pp 67–81

Foster RB Hubbell SP (1990a) Estructura de la vegetation y composición de especies de un lote de cincuenta hectáres en la isla de Barro Colorado. In: Leigh EG, Rand AS, Windsor DM (eds) Ecología de un bosque tropical. Ciclos estacionales y cambios a largo plazo. Smithsonian Institution Press, Washington, DC, pp 67–81

Foster RB, Hubbell SP (1990b) The floristic composition of the Barro Colorado Island forest. In: Gentry AH (ed) Four neotropical rainforests. Yale University Press, New Haven, pp 85–98

Fowler HG (1981) On the emigration of leaf-cutting ant colonies. Biotropica 13:316

Fowler HG (1983) Distribution patterns of Paraguayan leaf-cutting ants (Atta and Acromyrmex) (Formicidae: Attini). Stud Neotrop Fauna Environ 18:121–138

Fowler HG, Robinson SW (1979) Foraging by *Atta sexdens* (Formicida: Attini): seasonal patterns, caste and efficiency. Ecol Entomol 4:239–247

Fowler HG, Saes NB (1986) Dependence of the activity of grazing cattle on foraging grass-cutting ants (*Atta* spp.) in the southern Neotropics. J Appl Entomol 101:154–158

Fowler HG, Stiles EW (1980) Conservative resource management by leaf-cutting ants? The role of foraging territories and trails, and environmental patchiness. Sociobiology 5:25–41

Fowler HG, Forti LC, Pereira-da-Silva V, Saes NB (1986a) Economics of grass-cutting ants. In: Lofgren CS, Vander Meer RK (eds) Fire ants and leaf-cutting ants: biology and management. Westview Press, Boulder, CO, pp 18–35

Fowler HG, Pereira da Silva V, Forti LC, Saes NB (1986b) Population dynamics of leaf-cutting ants: a brief review. In: Lofgren CS, Vander Meer RK (eds) Fire ants and leaf-cutting ants: biology and management. Westview Press, Boulder, CO, pp 123–145

Fowler HG, Pagani MI, da Silva OA, Forti LC, Saes NB (1989) A pest is a pest is a pest? The dilemma of neotropical leaf-cutting ants: keystone taxa of natural ecosystems. Environ Manage 13:671–675

Fowler HG, Forti LC, di Romagnano LFT (1990) Methods for the evaluation of leaf-cutting ant harvest. In: Vander Meer RK, Jaffe K, Cedeno A (eds) Applied myrmecology – a world perspective. Westview Press, Boulder, CO, pp 228–241

Garrettson M, Stetzel JF, Halpern BS, Hearn DJ, Lucey BT, McKone MJ (1998) Diversity and abundance of understorey plants on active and abandoned nests of leaf-cutting ants (*Atta cephalotes*) in a Costa Rican rain forest. J Trop Ecol 14:17–26

Gentry AH (1982) Patterns of neotropical plant species diversity. Evol Biol 15:1–84

Gentry AH (1988) Changes in plant community diversity and floristic composition on environmental and geographical gradients. Ann Mo Bot Garden 75:1–34

Givnish TJ (1999) On the causes of gradients in tropical tree diversity. J Ecol 87:193–210

Glanz WE (1990) Neotropical mammal densities: how unusual is the community on Barro Colorado Island, Panama? In: Gentry AH (ed) Four neotropical rainforests. Yale University Press, New Haven, pp 287–311

Gombauld P, Rankin-de Merona J (1998) Influence of season on phenology and insect herbivory on saplings of tropical rain forest trees in French Guyana [French]. Ann Sci For 55:715–725

Gower ST, Norman JM (1991) Rapid estimation of leaf area index in conifer and broadleaf plantations. Ecology 72:1896–1900

Grace JC, Rook DA, Lane PM (1987) Modelling canopy photosynthesis in *Pinus radiata* stands. N Z J For Sci 17:204–228

Grantz DA, Zhang XJ, Metheney PD, Grimes DW (1993) Indirect measurement of leaf area index in Pima cotton (*Gossypium barbadense* L.) using a commercial gap inversion method. Agric For Meteorol 63:1–12

Grubb PJ (1995) Mineral nutrition and soil fertility in tropical rain forests. In: Lugo AE, Lowe C (eds) Tropical forests: management and ecology. Ecological studies, vol 112. Springer, Berlin Heidelberg New York, pp 308–330

Haines B (1975) Impact of leaf-cutting ants on vegetation development at Barro Colorado Island. In: Golley FG, Medina E (eds) Tropical ecological systems. Springer, Berlin Heidelberg New York pp 99–111

Haines B (1978) Element and energy flows through colonies of the leaf-cutting ant, *Atta colombica* in Panama. Biotropica 10:270–277

Hallé F, Oldman RAA, Tomlinson PB (1978) Tropical trees and forests. Springer Verlag, Berlin Heidelberg, New York

Haines BL (1983) Leaf-cutting ants bleed mineral elements out of a rain forest in southern Venezuela. Trop Ecol 24:85–93

Hamilton WD (1987) Kinship, recognition, disease, and intelligence: constraints of social evolution. In: Ito Y, Brow JL, Kikkawa J (eds) Animal societies: theories and facts. Japan Scientific Societies Press, Tokyo, pp 81–100

Hammond RB, Pedigo LP (1981) Effects of artificial and insect defoliation on water loss from excised soybean leaves. J Kansas Entomol Soc 54:331–336

Harley PC, Tenhunen JD, Lange OL (1986) Use of an analytical model to study limitations on net photosynthesis in *Arbutus unedo* under field conditions. Oecologia 70:393–401

Harley PC, Thomas RB, Reynolds JF, Strain BR (1992) Modelling photosynthesis of cotton grown in elevated CO_2. Plant Cell Environ 15:271–282

Hart AG, Ratnieks FLW (2001) Leaf caching in the leaf-cutting ant *Atta colombica*: organizational shift, task partitioning and making the best of a bad job. Anim Behav 62:227–234

Hart AG, Ratnieks FLW (2002) Waste management in the leaf-cutting ant *Atta colombica*. Behav Ecol 13:224–231

Haukioja E (1990) Induction of defence in trees. Annu Rev Entomol 36:25–42

Haukioja E, Honkanen T (1996) Why are tree responses to herbivory so variable. In: Mattson WJ, Niemelä P, Rousi M (eds) Dynamics of forest herbivory: quest for pattern and principle. USDA, For Serv Gen Tech Report, St Paul, pp 1–10

Herms DA, Mattson WJ (1992) The dilemma of plants: to grow or defend. Q Rev Biol 67:283–335

Hernández JV, Ramos C, Borjas M, Jaffe K (1999) Growth of *Atta laevigata* (Hymenoptera: Formicidae) nests in Pine plantations. Fla Entomol 82:97–103

Herz H (2001) Blattschneiderameisen im tropischen Regenwald. Freilanduntersuchungen zur Populationsdynamik sowie zu trophischen und nichttropischen Effekten von *Atta colombica* in Panama. PhD Diss, University of Bielefeld, Bielefeld, Germany

Hespenheide HA (1994) An overview of faunal studies. In: McDade LA, Bawa KS, Hespenheide HA, Hartshorn GS (eds) La Selva – ecology and natural history of a neotropical rain forest. The University of Chicago Press, Chicago, pp 238–243

Higashi S, Yamauchi K (1979) Influence of a supercolonial ant *Formica (Formica) yessensis* Forel on the distribution of other ants in Ishikari Coast. Jpn J Ecol 29:257–264

Hladik A, Hladik CM (1969) Rapports trophiques entre végétation et primates dans la forêt de Barro Colorado (Panama). La Terre et al Vie 23:25–117

Hodgson ES (1955) An ecological study of the behavior of the leaf-cutting ant *Atta cephalotes*. Ecology 36:293–304

Hölldobler B (1977) Communication in social Hymenoptera. In: Sebeok TA (ed) How animals communicate. Indiana University Press, Sunderland, MA

Hölldobler B, Lumsden CJ (1980) Territorial strategies in ants. Science 210:732–739

Hölldobler B, Roces F (2001) The behavioral ecology of stridulatory communication in leaf-cutting ants. In: Dugkatin LA (ed) Model systems in behavioral ecology: integrating empirical, theoretical and conceptual approaches. Princeton Univ Press, Princeton, NJ, pp 92–109

Hölldobler B, Wilson EO (1986) Nest area exploration and recognition in leafcutter ants (*Atta cephalotes*). J Insect Physiol 32:143–150

Hölldobler B, Wilson EO (1990) The ants. Harvard University Press, Cambridge, MA

Hölldobler B, Wilson EO (1994) Journey to the Ants. Harvard University Press, Cambridge, MA

Holbrook NM, Putz FE (1996) Physiology of tropical vines and hemiepiphytes: plants that climb up and plants that climb down. In: Mulkey SS, Chazdon RL, Smith AP (eds) Tropical forest plant ecophysiology. Chapman and Hall, New York, pp 363–394

Holdridge LR, Grenke WC, Hatheway WH, Liang T, Tosi JA Jr (1971) Forest environments in tropical live zones: a pilot study. Pergamon Press, Oxford

Holland EA, Parton WJ, Detling JK, Dyer MI (1992) Physiological responses of plant populations to herbivory and their consequences for ecosystem nutrient flow. Am Nat 140:685–706

Honkanen T, Haaukioja E, Suomela J (1994) Effects of simulated defoliation and debudding of needle and shoot growth in Scots Pine (*Pinus sylvestris*) – implication of plant source sink relationship for plant-herbivore studies. Funct Ecol 8:631–639

Horn HS (1971) The adaptive geometry of trees. Princeton University Press. Princeton, NJ

Horvitz CC, Schemske DW (1994) Effects of dispersers, gaps, and predators on dormancy and seedling emergence in a tropical herb. Ecology 75:1949–1958

Howard JJ (1987) Leaf-cutting ant diet selection: the role of nutrients, water, and secondary chemistry. Ecology 67:503–515

Howard JJ (1988) Leaf-cutting ant diet selection: relative influence of leaf-chemistry and physical factors. Ecology 69:250–260

Howard JJ (1990) Infidelity of leaf-cutting ants to host plants: resource heterogeneity or defense induction? Oecologia 82:394–401

Howard JJ (2001) Costs of trail construction and maintenance in the leaf-cutting ant *Atta colombica*. Behav Ecol Sociobiol 49:348–356

Howard JJ, Wiemer DF (1986) Chemical Ecology of host plant selection by the leaf-cutting ant, *Atta cephalotes*. In: Lofgren CS, Vander Meer RK (eds) Fire ants and leaf-cutting ants: biology and management. Westview Press, Boulder, CO, pp 260–273

Howard JJ, Green TP, Wiemer DF (1989) Comparative deterrency of two terpenoids to two genera of attine ants. J Chem Ecol 15:2279–2288

Howard JJ, Henneman L, Cronin G, Fox JA, Hormiga G (1996) Conditioning of scouts and recruits during foraging by a leaf-cutting ant *Atta colombica*. Anim Behav 52:299–306

Hubbell SP, Wiemer DF (1983) Host plant selection by an attine ant. In: Jaisson P (ed) Social insects in the tropics, vol 2. University of Paris Press, Paris pp 133–154

Hubbell SP, Johnson LK, Stanislaw E, Wilson B, Fowler H (1980) Foraging by bucket brigade in leaf-cutter ants. Biotropica 12:210–213

Hubbell SP, Wiemer DF, Adegare A (1983) An anti-fungal terpenoid defends a neotropical tree (Hymenaea) against attack by fungus-growing ants. Oecologia 60:321–327

Hubbell SP, Howard JJ, Wiener DF (1984) Chemical leaf repellency to an attine ant: seasonal distribution among potential host plant species. Ecology 65:1067–1076

Hubbell SP, Condit R, Foster RB (1990) Presence and absence of density dependence in a neotropical tree community. Philos Trans R Soc Lond Ser B 330:269–281

Hubbell SP, Foster RB, O'Brien ST, Harms KE, Condit R, Wechsler B, Wright SJ, Loo de Lao S (1999) Light-gap disturbances, recruitment limitation, and tree diversity in a neotropical forest. Science 283:554–557

Hulsmann J, Weissing FJ (1999) Biodiversity of plankton by species oscillations and chaos. Nature 402:407–410.

Huston M (1979) A general hypothesis of species diversity. Am Nat 113:81–101

Huston MA (1994) Biological diversity. The coexistance of species on changing landscapes. Cambridge University Press, Cambridge

Hutchings MJ, Wijesinghe DK, John EA (2000) The effects of heterogeneous nutrient supply on plant performance: a survey of responses, with special reference to clonal herbs. In: Hutchings MJ, John EA, Stewart AJA (eds) The Ecological consequences of environmental heterogeneity. The 40th Symposium of the British Ecological Society held at the University of Sussex, 23–25 March 1999. Blackwell Science, Oxford, pp 91–110

Isaaks E, Srivastava RM (1989) An introduction into applied geostatistics. Oxford University Press, New York

Ivlev V (1961) Experimental ecology of the feeding of fishes. Yale University Press, New Haven, CT

Jackson RB, Caldwell MM (1996) Integrating resource heterogeneity and plant plasticity: modelling nitrate and phosphate uptake in a patchy soil environment. J Ecol 84:891–903

Jaffe K (1986) Control of *Atta* and *Acromyrmex spp.* in pine tree plantations in the Venezuelan Llanos. In: Lofgren CS, Vander Meer RK (eds) Fire ants and leaf-cutting ants. Biology and management. Westview Press. Boulder, CO, pp 409–416

Jaffe K, Howse PE (1979) The mass recruitment system of the leaf-cutting ant *Atta cephalotes* (L). Anim Behav 27:930–939

Jaffe K, Vilela E (1989) On nest densities of the leaf-cutting ant *Atta cephalotes* in tropical primary forest. Biotropica 21:234–236

Janzen DH (1970) Herbivores and the number of tree species in tropical forests. Am Nat 104:501–528

Jentsch A (2001) The significance of disturbance for vegetation dynamics in sandy grassland ecosystems. PhD Diss, University of Bielefeld, Bielefeld, Germany

Johnson J (1999) Phylogenetic relationships within *Lepiota sensulato* based on morphological and molecular data. Mycologia 91:443–458

Jones CG, Lawton JH, Shachak M (1994) Organisms as ecosystem engineers. Oikos 69:373–386

Jonkman JCM (1980a) Average vegetative requirement, colony size and estimated impact of *Atta vollenweideri* on cattle-raising in Paraguay. Z Angew Entomol 89:135–143

Jonkman JCM (1980b) The external and internal structure and growth of nests of the leaf-cutting ant *Atta vollenweideri* Forel, 1893 (Hym.: Formicidae). Z Angew Entomol 89:158–173

Jordan CF, Uhl C (1978) Biomass of a "terra firme" forest of the Amazon basin. Oecol Plant 13:287–400

Kacelnik A (1993) Leaf-cutting ants tease optimal foraging theorists. Trends Ecol Evol 8:346–348

Karr JR (1990) The avifauna of Barro Colorado Island and the Pipeline Road, Panama. In: Gentry AH (ed) Four neotropical rainforests. Yale University Press, New Haven, CT, pp 183–198

Kaspari M (1993) Removal of seeds from Neotropical frugivore droppings: ant responses to seed number. Oecologia 95:81–88

Kaspari M (1996) Worker size and seed size selection by harvester ants in a neotropical forest. Oecologia 105:397–404

Kato R, Tadaki Y, Ogawa H (1978) Plant biomass and growth increment studies in Pasoh Forest. Malayan Nat J 30:211–224

Keller L, Genoud M (1997). Extraordinary lifespans in ants: a test of evolutionary theories of ageing. Nature 389:958–960

Kerr WE (1962) Tendências evolutivas na reproducao dos himenópteros sociais. Arq Mus Nacional Rio de Janeiro 52:115–116

Kira T (1978) Community architecture and organic matter dynamics in tropical lowland rain forests of southeast Asia with special reference to Pasoh Forest, West Malaysia. In: Tomlinson PB, Zimmermann MH (eds) Tropical trees as living systems. Cambridge University Press, Cambridge, pp 561–589

Kira T, Yoda K (1989) Vertical stratification in microclimate. In: Lieth H, Werger MJA (eds) Ecosystems of the world. Tropical rainforest ecosystems. Elsevier, Amsterdam, pp 55–71

Kira T, Shinozaki K, Hozumi K (1969) Structure of forest canopies as related to their primary productivity. Plant Cell Physiol 10:129–142

Kleineidam C, Roces F (2000) Carbon dioxide concentration and nest ventilation in nests of the leaf-cutting ant *Atta vollenweideri*. Insect Soc 47:241–248

Kleineidam C, Tautz J (1996) Perception of carbon dioxide and other "air-condition" parameters in the leaf-cutting ant *Atta cephalotes*. Naturwissenschaften 38:566–568

Kleineidam C, Ernst R, Roces F (2001) Wind-induced ventilation in the giant nests of the leaf-cutting ant *Atta vollenweideri*. Naturwissenschaften 88:301–305

Knapp JJ, Howse PE, Kermarrec A (1990) Foraging and fungal substrate selection by leaf-cutting ants. In: Vander Meer RK, Jaffe K, Cedeno A (eds) Applied myrmecology: a world perspective. Westview Press, Boulder, CO, pp 382–410

Knight DH (1963) A distance method for constructing forest profile diagrams and obtaining structural data. Trop Ecol 4:89–94

Köstner B, Lange OL (1986) Epiphytische Flechten in bayerischen Waldschadensgebieten des nördlichen Alpenraumes: floristisch-soziologische Untersuchungen und Vitalitätstests durch Photosynthesemessungen. Ber ANL 10:185–210

Kreisel H (1972) Pilze aus Pilzgärten von *Atta insularis* in Kuba. Z Allg Mikrobiol 12:643–654

Küppers M (1994) Canopy gaps: competitive light interception and economic space filling – a matter of whole-plant allocation. In: Caldwell MM, Pearcy RW (eds) Exploitation of environmental heterogeneity by plants. Academic Press, New York, pp 111–144

Landsberg J, Ohmart C (1989) Levels of insect defoliation in forests: patterns and concepts. Trends Ecol Evol 4:96–100

Lang ARG (1987) Simplified estimate of leaf area index from transmittance of the sun's beam. Agric For Meteorol 41:179–186

Lawton JH, Jones CG (1995) Linking species and ecosystems: organisms as ecosystem engineers. In: Jones CG, Lawton JH (eds) Linking species and ecosystems. Chapman and Hall, London, pp 141–150

Leal IR, Oliveira PS (1998) Interactions between fungus growing ants (Attini), fruits and seeds in Cerrado vegetation in Southeast Brazil. Biotropica 30:170–178

Leal IR, Oliveira PS (2000) Foraging ecology of attine ants in a neotropical savanna: seasonal use of fungal substrate in the cerrado vegetation of Brazil. Insect Soc 47:376–382

Lee DW (1987) The spectral distribution of radiation in two neotropical rainforests. Biotropica 19:161–166

Leigh EG (1999) Tropical forest ecology: a view from Barro Colorado Island. Oxford University Press, New York

Leigh EG, Smythe N (1978) Leaf production, leaf consumption and the regulation of folivory on Barro Colorado Island. In: Montgomery GG (ed) The ecology of arboreal folivores. Smithsonian Institution Press, Washington, DC, pp 33–50

Leigh EG, Windsor DM (1982) Forest production and regulation of primary consumers on Barro Colorado Island. In: Leigh EG Jr, Rand AS, Windsor DM (eds) The ecology of a tropical forest: seasonal rhythms and long term changes. Smithsonian Institute Press, Washington, DC, pp 111–122

Leigh EG, Wright SJ (1990) Barro Colorado Island and tropical biology. In: Gentry AH (ed) Four neotropical rainforests. Yale University Press, New Haven, CT, pp 28–47

Lemeur R (1973) A method for simulating the direct solar radiation regime in sunflower, Jerusalem artichoke, corn and soybean canopies using actual stand structure data. Agric Meteorol 12:229–247

Levey DJ, Byrne, MM (1993) Complex ant-plant interactions: rain forest ants as secondary dispersers and post-dispersal seed predators of rain forest plants. Ecology 74:1802–1812

Lewis T, Pollard GV, Dibley GC (1974a) Rhythmic foraging in the leaf cutting ant *Atta cephalotes* (L.) (Formicidae: Attini). J Anim Ecol 43:129–141

Lewis T, Pollard GV, Dibley GC (1974b) Microenvironmental factors affecting diel patterns of foraging in the leaf cutting ant *Atta cephalotes* (L.) (Formicidae: Attini). J Anim Ecol 43:143–153

Li-Cor Inc (1989) LAI-2000 plant canopy analyzer. Technical information. Li-Cor Inc, Lincoln, NE

Lieberman M, Lieberman D, Peralta R, Hartshorn GS (1995) Canopy closure and the distribution of tropical forest tree species at La Selva, Costa Rica. J Trop Ecol 11:161–178

Lighton JRB, Bartholomew GA, Feener DH (1987) Energetics of locomotion and load carriage and a model of the energy cost of foraging in the leaf-cutting ant *Atta colombica* Guer. Physiol Zool 60:524–537

Littledyke M, Cherrett JM (1976) Direct ingestion of plant sap from cut leaves by the leaf-cutting ants *Atta cephalotes* (L.) and *Acromyrmex octospinosus* (Reich) (Formicidae, Attini). Bull Entomol Res 66:205–217

Littledyke M, Cherrett JM (1978) Defence mechanisms in young and old leaves against cutting by the leaf-cutting ant *Atta cephalotes* (L). Bull Entomol Res 68:263–274

Lösch R, Schulze ED (1994) Internal coordination of plant responses to drought and evaporational demand. In: Schulze ED, Caldwell MM (eds) Ecophysiology of photosynthesis. Ecological studies, vol 100. Springer, Berlin Heidelberg New York, pp 185–204

Loreau M (1995) Consumers as maximizers of matter and energy flow in ecosystems. Am Nat 145:22–42

Loucks OL, Plumb-Mentjes ML, Rogers D (1985) Gap processes and large-scale disturbances in sand prairies. In: Pickett STA, White PS (eds) The ecology of natural disturbance and patch dynamics. Academic Press, San Diego, pp 71–83

Lowman MD (1984) An assessment of techniques for measuring herbivory: is rainforest defoliation more intense than we thought? Biotropica 16:264–268

Lowman MD (1985) Temporal and spatial variability in insect grazing of the canopies of five Australian rainforest tree species. Aust J Ecol 10:7–24

Lowman MD (1986) Light interception and its relation to structural differences in three Australian rainforest canopies. Aust J Ecol 11:163–170

Lowman MD (1995) Herbivory as a canopy process in rain forest trees. In: Lowman MD, Nadkarni NM (eds) Forest canopies. Academic Press, San Diego, pp 431–455

Lubchenco J (1978) Plant species diversity in a marine intertidal community: importance of herbivore food preference and algal competitive abilities. Am Nat 112:23–39

Lugo AE, Farnworth EG, Pool G, Jerez P, Kaufmann G (1973) The impact of the leaf-cutter ant *Atta colombica* on the energy flow of a tropical wet forest. Ecology 54:1292–1301

Lusk CH, Smith B (1998) Life history differences and tree species coexistence in an old-growth New Zealand rain forest. Ecology 79:795–806

Lutz FE (1929) Observations on leaf-cutting ants. Am Mus Novitates 388:1–21

MacArthur RH (1969) Patterns of communities in the tropics. Biol J Linn Soc 1:19–30

Mackey RL, Currie DJ (2000) A re-examination of the expected effects of disturbance on diversity. Oikos 88:483–493

Markl H (1965) Stridulation in leaf-cutting ants. Science 149:1392–1393

Markl H (1968) Die Verständigung durch Stridulationssignale bei Blattschneiderameisen. II: Erzeugung und Eigenschaften der Signale. Z Vergl Physiol 60:103–150

Marks PL (1974) The role of pin cherry (*Prunus pensylvanica* L.) in the maintenance of stability in northern hardwood ecosystems. Ecol Monogr 44:73–88

Marquis RJ (1984) Leaf herbivory decreases fitness of a tropical plant. Science 226:537–539

Marquis RJ (1992a) A bite is a bite is a bite? Constraints on response to folivory in *Piper arieianum* (Piperaceae). Ecology 73:143–152

Marquis RJ (1992b) Selective impact of herbivores. In: Fritz RS, Simms EL (eds) Ecology and evolution of plant resistance. Univ of Chicago Press, Chicago, pp 301–325

Marquis RJ, Braker HE (1994) Plant-herbivore interactions: diversity, specificity, and impact. In: McDade LA, Bawa KS, Hespenheide HA, Hartshorn GS (eds) La Selva – ecology and natural history of a neotropical rain forest. University of Chicago Press, Chicago, pp 261–281

Marquis RJ, Miller Alexander H (1992) Evolution of resistance and virulence in plant-herbivore and plant-pathogen interactions. Trends Ecol Evol 7:126–129

Marshall LG, Butler RE, Drake RE, Curtis GH, Tedford RH (1979) Calibration of the great American interchange. Science 204:272–279

Martin MM (1970) The biochemical basis of the fungus-attine ant symbiosis. Science 169:16–20

Martin MM, Weber NA (1969) The cellulose utilising capacity of the fungus cultured by the attine ant *Atta colombica tonsipes*. Ann Entomol Soc Am 62:11–13

Maschwitz U (1974) Vergleichende Untersuchungen zur Funktion der Ameisenmetathoracaldrüse. Oecologia 16:303–310

Maschwitz U, Hänel H (1985) The migrating herdsman *Dolichoderus (Diabolus) cuspidatus*: an ant with a novel mode of life. Behav Ecol Sociobiol 17:171–184

Maschwitz U, Koob K, Schildknecht H (1970) Ein Beitrag zur Funktion der Metathoracaldrüse der Ameisen. J Insect Physiol 16:387–404

Mathews JNA (1994) The benefits of overcompensation and herbivory: the differences between coping with herbivores and liking them. Am Nat 144:528–533

Mattson WJ, Addy ND (1975) Phytophagous insects as regulators of forest primary production. Science 190:515–522

Mazancourt de C, Loreau M, Abbadie L (1998) Grazing optimisation and nutrient cycling: when do herbivores enhance plant production. Ecology 79:2242–2252

McBrien H, Harmson R, Crowder A (1983) A case of insect grazing affecting plant succession. Ecology 64:1035–1039

McMillen GG, McClendon JH (1979) Leaf angle: an adaptive feature of sun and shade leaves. Bot Gaz 140:437–442

McNaughton SJ (1987) On plants and herbivores. Am Nat 128:765–770

McNaughton SJ (1979) Grazing as an optimization process: grass-ungulate relationships in the Serengeti. Am Nat 113:691–703

McNaughton SJ, Oesterheld M, Frank DA, Williams KJ (1989) Ecosystems-level patterns of primary productivity and herbivory in terrestrial habitats. Nature 341:142–144

McWilliam A-L, Roberts JM, Cabral OMR, Leitao MVBR, de Costa ACL, Roberts JM, Zamparoni CAGP (1993) Leaf area index and above ground biomass of terra firme rain forest and adjacent clearings in Amazonia. Funct Ecol 7:310–317

Midgley JJ, Cameron MC, Bond WJ (1995) Gap characteristics and replacement patterns in the Knysna Forest, South Africa. J Veg Sci 6:29–36

Milhahn K (2001) Quantitative Analyse der durch "Wasting" von Mantelbrüllaffen (*Alouatta palliata*) verursachten Blattverluste in einem tropischen halbimmergrünen Regenwald in Panama. Diploma Thesis, University of Bielefeld, Bielefeld, Germany

Mitchell PL, Whitmore TC (1993) Use of hemisperical photographs in forest ecology. OFI Occasional Papers 44. Oxford Forestry Institute, Oxford

Mole S, Ross JAM, Waterman PG (1988). Light induced variation in phenolic levels in foliage of rain-forest plants. I. Chemical changes. J Chem Ecol 14:1–21

Monsi M, Murata Y (1970) Development of photosynthetic systems as influenced by distribution of matter. In: Setlik I (ed) Prediction and measurement of photosynthetic productivity. PUDOC, Wageningen, pp 115–130

Monsi M, Saeki T (1953) Über den Lichtfaktor in den Pflanzengesellschaften und seine Bedeutung für die Stoffproduktion. Jpn J Bot 14:22–52

Montgomery RA, Chazdon RL (2002) Light gradient partitioning by tropical tree seedlings in the absence of canopy gaps. Oecologia 131:165–174

Morini MSDC, Fowler HG, Schlittler FM, Bueno OC (1993) Polymorphic and relative pick-up responses of a leaf-cutting ant, *Atta sexdens rubropilosa* (Hymenoptera: Formicidae), to extracts from new and old *Eucalyptus* leaves. J Plant Protect Trop 10:59–62

Moser JC (1963) Contents and structure of *Atta texana* nest in summer. Ann Entomol Soc Am 56:286–291

Moser JC (1967) Mating activities of *Atta texana* (Hymenoptera, Formicidae). Insect Soc 14:295–312

Moser J, Blum MS (1963) Trail marking substance of the Texas leaf-cutting ant: source and potency. Science 140:1228

Moutinho PRS, Nepstad DC, Araujo K, Uhl C (1993) Formigas e floresta: estudo para a recuperaçao de areas de pastagem. Ciencia Hoje 88: 59–60

Müller D, Nielsen J (1965) Production brute, pertes par respiration et production nette dans la forêt ombrophile tropicale. For Forsoegsvaes Dan 29:69–160

Mueller UG, Rehner SA, Schlutz TR (1998) The evolution of agriculture in ants. Science 281:2034–2038

Mueller UG, Schultz TR, Currie CR, Adams RMM, Malloch D (2001) The origin of the attine ant-fungus mutualism. Q Rev Biol 76:169–197

Mulkey SS, Wright SJ (1996) Influence of seasonal drought on the carbon balance of tropical forest plants. In: Mulkey SS, Chazdon RL, Smith AP (eds) Tropical forest plant ecophysiology. Chapman and Hall, New York, pp 187–216

Murphy PG, Lugo AE (1986) Structure and biomass of a subtropical dry forest in Puerto Rico. Biotropica 18:89–96

Nepstad D, Uhl C, Serrao E (1990) Surmounting barriers to forest regeneration in abandoned highly degraded pastures: a case study from Paragominas, Para, Brazil. In: Anderson A (ed) Alternatives to deforestation, steps towards sustainable use of the Amazon rain forest. Columbia University Press, New York, pp 215–229

Newbery DM, De Foresta H (1985) Herbivory and defense in pioneer, gap and understory trees of tropical rain forest in French Guiana. Biotropica 17:238–244

Nichols-Orians CM (1982) The acceptability of young and mature leaves to leaf-cutter ants varies with light environment. Biotropica 24:211-214

Nichols-Orians CM (1991a) Condensed tannins, attine ants, and the performance of symbiontic fungus. J Chem Ecol 17:1177–1195

Nichols-Orians CM (1991b) The effects of light on foliar chemistry, growth and susceptibility of seedlings of a canopy tree to an attine ant. Oecologia 86:552–560

Nichols-Orians CM (1991 c) Environmentally induced differences in plant traits: consequences for susceptibility to a leaf-cutter ant. Ecology 72:1609–1623

Nichols-Orians CM, Schultz JC (1990) Interactions among leaf toughness, chemistry, and harvesting by attine ants. Ecol Entomol 15:311–320

Noble IR, Slatyer RO (1980) The use of vital attributes to predict successional changes in plant communities subject to recurrent disturbances. Vegetatio 43:5–21

Norman JM, Campbell GS (1989) Canopy structure. In: Pearcy RW, Ehleringer J, Mooney HA, Rundel PW (eds) Plant physiological ecology: field methods and instrumentation. Chapman and Hall, London, pp 301–325

Norris DM (1988) Sensitivity of insect-damaged plants to environmental stresses. In: Heinrichs EA (ed) Plant stress – insect interactions. Wiley, New York, pp 341–361

North RD, Jackson CW, Howse PE (1997) Evolutionary aspects of ant-fungus interactions in leaf-cutting ants. Trends Ecol Evol 12:386–389

North RD, Jackson CW, Howse PE (1999) Communication between the fungus garden and workers of the leaf-cutting ant, Atta sexdens rubropilosa, regarding choice of substrate for the fungus. Physiol Entomol 24:127–133

Novotny V, Basset Y, Miller SE, Weiblen GD, Bremer B, Cizek L, Drozd P (2002) Low host specifity of herbivorous insects in a tropical forest. Nature 416:841–844

Oba G (1994): Responses of Indigofera spinosa to simulated herbivory in a semidesert of north-west Kenya. Acta Oecol 15:105–117

Oberbauer SF, Donelly MA (1986) Growth analysis and successional status of Costa Rican rain forest trees. New Phytol 104:517–521

Odum HT, Ruiz-Reyes J (1970) Holes in leaves and the grazing control mechanism. In: Odum HT, Pigeon RF (eds) A tropical rain forest. US Atomic Energy Commission, Oak Ridge, TE, pp I-69-I-80

Oesterheld M, McNaughton SJ (1991) Effects of stress and time for recovery on the amount of compensatory growth after grazing. Oecologia 85:305–313

Oliveira PS, Galetti M, Pedroni F, Morellato FPC (1995) Seed cleaning by *Mycocepurus goeldii* ants (Attini) facilitates germination in *Hymenaea courbaril* (Caesalpinaceae). Biotropica 27:518–522

Orr MR (1992) Parasitic flies (Diptera: Phoridae) influence foraging rhythms and caste division of labor in leaf-cutter ant, *Atta cephalotes* (Hymenoptera: Formicidae). Behav Ecol Sociobiol 30:395–402

Ortius-Lechner D, Maile R, Morgan ED, Boomsma JJ (2000) Metapleural gland secretions of the leaf-cutter ant *Acromyrmex octospinosus*: new compounds and their functional significance. J Chem Ecol 26:1667–1683

Ostlie KR, Pedigo LP (1984) Water loss from soybeans after simulated and actual insect defoliation. Environ Entomol 13:1675–1680

Otto H (1994) Waldökologie. (UTB). Ulmer Verlag, Stuttgart

Ovaska J, Wallis M, Mutikainen P (1992) Changes in leaf gas exchange properties of cloned *Betula pendula* saplings after partial defoliation. J Exp Bot 43:1301–1307

Owen DF, Wiegert RG (1976) Do consumers maximize plant fitness? Oikos 27:488–492

Pacala SW, Crawley MJ (1992) Herbivores and plant diversity. Am Nat 140:243–260

Parker GG (1995) Structure and microclimate of forest canopies. In: Lowman MD, Nadkarni NM (eds) Forest canopies. Academic Press, San Diego, pp 73–106

Passos L, Ferreira SO (1996) Ant dispersal of *Croton priscus* (Euphorbiaceae) seeds in a tropical semideciduous forest in southeastern Brazil. Biotropica 28:697–700

Pearcy RW (1983) The light environment and growth of C3 and C4 tree species in the understory of a Hawaiian forest. Oecologia 58:19–25

Pearcy RW, Sims DA (1994) Photosynthetic acclimation to changing light environments: scaling from the leaf to the whole plant. In: Caldwell MM, Pearcy RW (eds) Exploitation of environmental heterogeneity by plants. Academic Press, New York, pp 145–174

Pearcy RW, Chazdon RL, Gross LJ, Mott KA (1994) Photosynthetic utilization of sunflecks: a temporally patchy resource on a time scale of seconds to minutes. In: Caldwell MM, Pearcy RW (eds) Exploitation of environmental heterogeneity by plants. Academic Press, New York, pp 175–208

Perfecto I, Vandermeer J (1993) Distribution and turnover rate of a population of *"Atta cephalotes"* in a tropical rain forest in Costa Rica. Biotropica 25:316–321

Perry DA (1994) Forest ecosystems. Johns Hopkins University Press, Baltimore

Pfitsch WA, Pearcy RW (1992) Growth and reproduction allocation of *Adenocaulon bicolor* following the experimental removal of sunflecks. Ecology 73:2109–2117

Pickett STA, White PS (1985) (eds) The ecology of natural disturbance and patch dynamics. Academic Press, Orlando, pp 3–13

Pickett STA, Cadenasso ML, Jones CG (2000) Generation of heterogeneity by organism: creation, maintenance and transformation. In: Hutchings MJ, John EA, Stewart AJA (eds) The ecological consequences of environmental heterogeneity. The 40th Symposium of the British Ecological Society held at the University of Sussex, 23–25 March 1999. Blackwell Science, Oxford, pp 33–52

Pilati A, Quiran E (1996) Foraging patterns of *Acromyrmex lobicornis* (Formicidae: Attini) in a grassland of Lihue Calel national Park, La Pampa, Argentina. Ecol Aust 6:123–126

Pizo MA, Oliveira PS (1998) Interaction between ants and seeds of a nonmyrmecochorous neotropical tree, *Cabralea canjerana* (Meliaceae), in the atlantic forest of southeast Brazil. Am J Bot 85:669–674

Platt WJ (1975) The colonization and formation of equilibrium plant species associations on badger disturbances in a tall-grass prairie. Ecol Monogr 45:285–305

Pons TL (1989) Breaking of seed dormancy by nitrate as a gap detection mechanism. Ann Bot 63:139–143

Porter SD, Bowers MA (1980) Emigration of an *Atta* colony. Biotropica 12:232–233

Powell RJ, Stradling DJ (1986) Factors influencing the growth of *Attamyces bromatificus* a symbiont of attine ants. Trans Br Mycol Soc 87:205–213

Putz FE (1983) Liana biomass and leaf area of a "tierra firme" forest in the Rio Negro basin, Venezuela. Biotropica 15:185–189

Putz FE (1988) Woody vines and tropical forests. Fairchild Tropical Gard Bull 43:5–13

Pyke G (1984) Optimal foraging theory: a critical review. Annu Rev Ecol Syst 15:523–575

Pyke GH, Pulliam HR, Charnov GL (1977) Optimal foraging: a selective review of theory and tests. Q Rev Biol 52:137–154

Quinland RJ, Cherrett JM (1978) Aspects of the symbiosis of the leaf-cutting ant *Acromyrmex octospinosus* (Reich) and its food fungus. Ecol Entomol 3:221–230

Quinland R, Cherrett JM (1979) The role of fungus in the diet of the leaf-cutting ant *Atta cephalotes* (L.). Ecol Entomol 4:151–160

Rand AS, Myers CW (1990) The herpetofauna of Barro Colorado Island, Panama: an ecological summary. In: Gentry AH (ed) Four neotropical rainforests. Yale University Press, New Haven, pp 386–409

Rankin-de-Morena JM, Hutchings HRW, Lovejoy TE (1990) Tree mortality and recruitment over a five year period in undisturbed upland rain forest of the central Amazon. In: Gentry AH (ed) Four neotropical rainforests. Yale Univ Press, New Haven, pp 573–584

Rao M (2000) Variation in leaf-cutter ant (*Atta* sp.) densities in forest isolates: the potential role of predation. J Trop Ecol 16:209–225

Rao M, Terborgh J, Nuñez P (2001) Increased herbivory in forest isolates: implications for plant community structure and composition. Conserv Biol 15:624–633

Reed J, Cherrett JM (1990) Foraging strategies and vegetation exploitation in the leaf-cutting ant *Atta cephalotes* (L.) – a preliminary simulation model. In: van der Meer RK, Jaffe K, Cedeno A (eds) Applied myrmecology – a world perspective, Westview Press, Boulder, CO, pp 355–366

Rhoades DF, Cates RG (1976) Toward a general theory of plant antiherbivore chemistry. Rec Adv Phytochem 10:168–213

Ribeiro GT, Woessner RA (1979) Teste de eficiencia com seis sauvicidas no controle de sauvas (*Atta spp.*) na Jari, Pará, Brasil. An Soc Entomol Brasil 8:77–84

Rich MP, Clark DB, Clark DA, Oberbauer SF (1993) Long-term study of solar radiation regimes in a tropical wet forest using quantum sensors and hemispherical photography. Agric For Meteorol 65:107–127

Ricklefs RE (1977) Environmental heterogeneity and plant species diversity: a hypothesis. Am Nat 111:376–381

Ridley PS, Howse PE, Jackson CW (1996) Control of the behaviour of leaf-cutting ants by their 'symbiotic' fungus. Experientia 52:631–635

Roberts JM, Cabral OMR, Fisch G, Molion LCB, Moore CJ, Shuttleworth WJ (1993) Transpiration from an Amazonian rainforest calculated from stomatal conductance measurements. Agric For Meteorol 62:175–196

Roberts JT, Heithaus ER (1986) Ants rearrange the vertebrate-generated seed shadow of a neotropical fig tree. Ecology 67:1046–1051

Roces F (1990) Olfactory conditioning during the recruitment process in a leaf-cutting ant. Oecologia 83:261–262

Roces F (1994) Odor learning and decision making during food collection in the leaf-cutting ant *Acromyrmex lundi*. Insect Soc 41:235–239

Roces F, Hölldobler B (1994) Leaf density and a trade-off between load-size selection and recruitment behavior in the ant *Atta cephalotes*. Oecologia 97:1–8

Roces F, Hölldobler B (1995) Vibrational communication between hitchhikers and foragers in leaf-cutting ants (*Atta cephalotes*). Behav Ecol Sociobiol 37:297–302

Roces F, Hölldobler B (1996) Use of stridulation in foraging leaf-cutting ants: mechanical support during cutting or short-range recruitment signal? Behav Ecol Sociobiol 39:293–299

Roces F, Lighton JRB (1995) Larger bites of leaf-cutting ants. Nature 373:392–393

Roces F, Núñez JA (1993) Information about food quality influences load-size selection in recruited leaf-cutting ants. Anim Behav 45:135–143

Roces F, Tautz J, Hölldobler B (1993) Stridulation in leaf-cutting ants: short-range recruitment through plant-borne vibrations. Naturwissenschaften 80:521–524

Rockwood LL (1973) Distribution, density, and dispersion of two species of Atta (Hymenoptera: Formicidae) in Guanacaste Province, Costa Rica. J Anim Ecol 42:803–817

Rockwood LL (1975) The effects of seasonality on foraging in two species of leaf-cutting ants (Atta) in Guanacaste Province, Costa Rica. Biotropica 7:176–193

Rockwood LL (1976) Plant selection and foraging patterns in two species of leaf-cutting ants (Atta). Ecology 57:48–61

Rockwood LL, Hubbell SP (1987) Host plant selection, diet, diversity, and optimal foraging in a tropical leaf-cutting ant. Oecologia 74:55–61

Röschard J, Roces F (2002) The effect of load length, width and mass on transport rate in the grass-cutting ant Atta vollenweideri. Oecologia 131:319–324

Rojas P (1989) Entomofauna associated with the detritus of Atta mexicana F. Smith [Hymenoptera: Formicidae] in an arid zone of central Mexico. Acta Zool Mex Nueva Ser 33:1–52

Ross J (1981) The radiation regime and architecture of plant stands. Junk, The Hague

Rudolph SG, Loudon C (1986) Load-size selection by foraging leaf-cutter ants (Atta cephalotes). Ecol Entomol 11:401–410

Ryel RJ, Beyschlag W (2000) Gap dynamics. In: Marshall B, Roberts J (eds) Leaf development and canopy growth. Sheffield biological science series. Sheffield Academic Press, Sheffield, pp 251–279

Ryel RJ, Caldwell MM (1998) Nutrient acquisition from soils with patchy nutrient distributions: importance of patch size, degree of variability and root uptake kinetics. Ecology 79:2735–2744

Ryel RJ, Barnes PW, Beyschlag W, Caldwell MM, Flint SD (1990) Plant competition for light analyzed with a multispecies canopy model. I. Model development and influence of enhanced UV-B conditions on photosynthesis in mixed wheat and wild oat canopies. Oecologia 82:304–310

Ryel RJ, Beyschlag W, Caldwell MM (1994) Light field heterogeneity among tussock grasses: theoretical considerations of light harvesting and seedling establishment. Oecologia 98:241–246

Ryel RJ, Beyschlag W, Heindl B, Ullmann I, Klein M (1996) Experimental studies on the competitive balance between two Central European roadside grasses with different growth forms: 1. Field experiments on the effects of mowing and maximum leaf temperatures on competitive ability. Bot Acta 109:441–448

Ryel RJ, Falge E, Joss U, Geyer R, Tenhunen JD (2001) Penumbral and foliage distribution effects on Pinus sylvestris canopy gas exchange. Theor Appl Climatol 68:109–124

Salatino A, Sugayama RL, Negri G, Vilegas W (1998) Effect of constituents of the foliar wax of Didymopanax vinosum on the foraging activity of the leaf-cutting ant Atta sexdens rubropilosa. Entomol Exp Appl 86:261–266

Salzemann A, Jaffe K (1990a) Territorial ecology of the leaf-cutting ant, Atta laevigata. In: Vander Meer RK, Jaffe K, Cedeno A (eds) Applied myrmecology. A world perspective. Westview Press, Boulder, CO, pp 345–354

Salzemann A, Jaffe K (1990b) On the territorial behaviour of field colonies of the leaf-cutting ant Atta laevigata (Hymenoptera: Myrmicinae). J Insect Physiol 36:133–138

Schaefer M (1992) Wörterbücher der Biologie. Ökologie. UTB, Gustav Fischer Verlag, Jena

Schierenbeck KA, Mack RN, Sharitz RR (1994) Effects of herbivory on growth and bio-mass allocation in native and introduced species of *Lonicera*. Ecology 75:1661–1672

Schildknecht H, Koob K (1970) Plant bioregulators in the metathoracic glands of myr-micine ants. Angew Chem (Int Edn) 9:173

Schildknecht H, Koob K (1971) Myrmicacin, the first insect herbicide. Angew Chem (Int Edn) 10:124–125

Schlensog M (1997) Experimentelle Untersuchungen des Lichtklimas in Urwald-parzellen Nordborneos. Göttinger Beiträge zur Land- und Forstwirtschaft in den Tropen und Subtropen, Bd 117. Verlag Erich Golze, Göttingen

Schneirla TC (1933) Study on army ants in Panama. J Comp Psychol 25:51–90

Schultz TR, Meier R (1995) A phylogenetic analysis of the fungus-growing ant (Hymenoptera: Formicidae: Attini) based on morphological characters of the larvae. Syst Entomol 20:337–370

Shepherd JD (1982) Trunk trails and the searching strategy of a leaf-cutter ant, *Atta colombica*. Behav Ecol Sociobiol 11:77–84

Shepherd JD (1985) Adjusting foraging effort to resources in adjacent colonies of the leaf-cutting ant *Atta colombica*. Biotropica 17:245–252

Sherman PW, Seeley TD, Reeve HK (1988) Parasites, pathogens, and polyandry in social hymenoptera. Am Nat 131:602–610

Simpson EH (1949) Measurement of diversity. Nature 163:668

Siqueira de CG, Bacci M Jr, Pagnocca FC, Bueno OC, Hebling MJA (1998) Metabolism on plant polysaccharides by *Leucoagaricus gongylophorus*, the symbiotic fungus of the leaf-cutting ant *Atta sexdens* L. Appl Environ Microbiol 64:4820–4822

Smith AP, Hogan KP, Idol JR (1992) Spatial and temporal patterns of light and canopy structure in a lowland tropical moist forest. Biotropica 24:503–511

Smith FW, Sampson DA, Long JN (1991) Comparison of leaf area index estimates from tree allometrics and measured light interception. For Sci 37:1682–1688

Smith WK, Knapp AK, Reiners WA (1989) Penumbral effects on sun light penetration in plant communities. Ecology 70:1603–1609

Solé RV, Manrubia SC (1995) Are rainforests self-organized in a critical state? J Theor Biol 173:31–40

Sperry JS, Adler FR, Campbell GS, Comstock JP (1998) Limitation of plant water use by rhizosphere and xylem conductance: results from a model. Plant Cell Environ 21:347–359

Stahel G, Geijskes DC (1939) Über den Bau der Nester von *Atta cephalotes* L. und *Atta sexdens* L. (Hym. Formicidae). Rev Entomol 10:27–78

Sterck F, Vander Meer P, Bongers F (1992) Herbivory in two rainforest canopies in French Guyana. Biotropica 24:97–99

Stewart AJA, John EA, Hutchings MJ (2000) The world is heterogeneous: ecological con-sequences of living in a patchy environment. In: Hutchings MJ, John EA, Stewart AJA (eds) The ecological consequences of environmental heterogeneity. The 40th Sympo-sium of the British Ecological Society held at the University of Sussex, 23–25 March 1999. Blackwell Science, Oxford, pp 1–8

Swift MJ, Heal OW, Anderson JM (1979) Decomposition in terrestrial ecosystems. Uni-versity of California Press, Berkeley

Tautz J, Roces F, Hölldobler B (1995) Use of sound-based vibratome by leaf-cutting ants. Science 267:84–87

Tenhunen JD, Harley PC, Beyschlag W, Lange OL (1987) A model of net photosynthesis for leaves of the sclerophyll *Quercus coccifera*. In: Tenhunen JD, Catarino F, Lange OL, Oechel WC (eds) Plant response to stress – functional analysis in mediterranean ecosystems. Springer, Berlin Heidelberg New York pp 339–354

Tenhunen JD, Sala Serra A, Harley PC, Dougherty RL, Reynolds JR (1990) Factors influencing carbon fixation and water use by Mediterranean sclerophyll shrubs during summer drought. Oecologia 82:381–393

Tenhunen JD, Siegwolf RA, Oberbauer SF (1994) Effects of phenology, physiology and gradient in community composition, structure and microclimate on tundra ecosystem CO_2 exchange. In: Schulze ED, Caldwell MM (eds) Ecophysiology of photosynthesis. Ecological studies, vol 100. Springer, Berlin Heidelberg New York, pp 431–460

Thorington RW Jr, Tannenbaum B, Tarak A, Rudran R (1996) Distribution of trees on Barro Colorado Island: a five-hectare sample. In: Leigh EG Jr, Rand AS, Windsor DM (eds) The ecology of a tropical forest. Seasonal rhythms and long-term changes. Smithsonian Institution Press, Washington, DC, pp 83–94

Tilman D (1982) Resource competition and community structure. Princeton University Press, Princeton, NJ

Tilman D (1988) Plant strategies and the dynamics and structure of plant communities. Princeton University Press, Princeton, NJ, USA

Tobin MF, Lopez OR, Kursar TA (1999) Responses of tropical understory plants to a severe drought: tolerance and avoidance of water stress. Biotropica 31:570–578

Tonhasca A Jr (1996) Interactions between a parasitic fly, *Neodohrniphora declinata* (Dipera:Phoridae), and its host, the leaf-cutting ant *Atta sexdens rubropilosa* (Hymenoptera: Formicidae). Ecotropica 2:157–164

Tonhasca A Jr, Braganca MAL (2000) Effect of leaf toughness on the susceptibility of the leaf-cutting ant *Atta sexdens* to attacks of a phorid parasitoid. Insect Soc 47:220–222

Torquebiau EF (1988) PAR environment patch dynamics and architecture in a tropical rainforest in Sumatra. Aust J Plant Physiol 15:327–342

Trumble JT, Kolodny-Hirsch DM, Ting IP (1993) Plant compensation for arthropod herbivory. Annu Rev Entomol 38:93–119

Tumlinson JH, Silverstein RM, Moser JC (1971) Identification of the trail pheromone of a leaf-cutting ant *Atta texana*. Nature 234:348–349

Tumlinson JH, Moser JC, Silverstein RM (1972) A volatile trail pheromone of the ant *Atta texana*. J Insect Physiol 18:809–814

Tuomi J, Nilsson P, Astrom M (1994) Plant compensatory responses: bud dormancy as an adaptation to herbivory. Ecology 75:1429–1436

Tyree MT, Sperry JS (1988) Do woody plants operate near the point of catastrophic xylem dysfunction caused by dynamic water stress? Plant Physiol 88:574–580

Van der Maarel E (1993) Some remarks on disturbance and its relations to diversity and stability. J Veg Sci 4:733–736

Vander Meer RK, Jaffe K, Cedeno A (eds) (1990) Applied myrmecology – a world perspective. Westview Press, Boulder, CO

Vasconcelos HL (1988) Distribution of *Atta* (Hymenoptera – Formicidae) in "terrafirme" rain forest of Central Amazonia: density, species composition and preliminary results on effects of forest fragmentation. Acta Amazon 18:309–315

Vasconcelos HL (1990a) Habitat selection by the queens of the leaf-cutting ant *Atta sexdens* L. in Brazil. J Trop Ecol 6:249–252

Vasconcelos HL (1990b) Foraging activity of two species of leaf-cutting ants (*Atta*) in a primary forest of the central amazon. Insect Soc 37:131–146

Vasconcelos HL (1997) Foraging activity of an Amazonian leaf-cutting ant: responses to changes in the availability of woody plants and to previous plant damage. Oecol Berl 112:370–373

Vasconcelos HL (1999) Levels of leaf herbivory in Amazonian trees from different stages in forest regeneration. Acta Amazon 29:615–623

Vasconcelos HL, Casimiro AB (1997) Influence of *Azteca alfari* ants on the exploitation of *Cecropia* trees by a leaf-cutting ant. Biotropica 29:84–92

Vasconcelos HL, Cherrett JM (1995) Changes in leaf-cutting ant populations (Formicidae: Attini) after the clearing of mature forest in Brazilian Amazonia. Stud Neotrop Fauna Environ 30:107–113

Vasconcelos HL, Cherrett HG (1996) The effect of wilting on the selection of leaves by the leaf-cutting ant *Atta laevigata*. Entomol Exp Appl 78:215–220

Vasconcelos HL, Cherrett JM (1997) Leaf-cutting ants and early forest regeneration in central Amazonia: effects of herbivory on tree seedling establishment. J Trop Ecol 13:357–370

Vasconcelos HL, Fowler HG (1990) Foraging and fungal substrate selection by leaf-cutting ants. In: Vander Meer RK, Jaffe K, Cedeno A (eds) Applied myrmecology – a world perspective. Westview Press, Boulder, CO, pp 411–419

Vázquez-Yanes C, Orozco-Segovia A (1994) Signals for seeds to sense and respond to gaps. In: Caldwell MM, Pearcy RW (eds) Exploitation of environmental heterogeneity by plants. Academic Press, San Diego, pp 209–236

Vilela EF, Howse PE (1986) Territoriality in leaf-cutting ants, *Atta* spp. In: Lofgren CS, Vander Meer RK (eds) Fire ants and leaf-cutting ants. Biology and management. Westview Press. Boulder, CO, pp 159–171

Vitousek PM, Sanford RL (1986) Nutrient cycling in moist tropical forest. Ann Rev Ecol Syst 17:137–167

Vogl RJ (1974) Effects of fire on grasslands. In: Kozlowski TT, Ahlgren CE (eds) Fire and ecosystems. Academic Press, New York, pp 139–194

Walker B, Kinzig A, Langridge J (1999) Plant attributes diversity, resilience, and ecosystem function: the nature and significance of dominant and minor species. Ecosystems 2:95–113

Wall DH, Moore JC (1999) Interactions underground. BioScience 49:109–117

Waller DA (1982) Leaf-cutting ants and live oak: the role of leaf toughness in seasonal and intraspecific host choice. Entomol Exp Appl 32:146–150

Waller DA (1986) The foraging ecology of *Atta texana* in Texas. In: Lofgren CS, Vander Meer RK (eds) Fire ants and leaf-cutting ants. Westview Press, Boulder, CO, pp 146–158

Waller DA, Moser JC (1990) Invertebrate enemies and nest associates of the leaf-cutting ant *Atta texana* (Buckley) (Formicidae, Attini). In: Vander Meer RK, Jaffe K, Cedeno A (eds) Applied myrmecology – a world perspective. Westview Press, Boulder, CO, pp 255–273

Watling JR, Press MC (2000) Light heterogeneity in tropical rain forests: photosynthetic responses and their ecological consequences. In: Hutchings MJ, John EA, Stewart AJA (eds) The ecological consequences of environmental heterogeneity. The 40th Symposium of the British Ecological Society held at the University of Sussex, 23–25 March 1999. Blackwell Science, Oxford, pp 131–153

Weber NA (1941) The biology of the fungus-growing ants. Part VII. The Barro Colorado Island, canal zone, species. Rev Entomol 12:93–131

Weber NA (1966) Fungus growing ants. Science 153:587–604

Weber NA (1969) Ecological relations of three *Atta* species in Panama. Ecology 50:141–147

Weber NA (1972a) The fungus-culturing behavior of ants. Am Zool 12:577–587

Weber NA (1972b) Gardening ants: the attines. Am Philos Soc, Philadelphia, PA

Weber NA (1979) Fungus culturing by ants. In: Batra LR (ed) Insect-fungus symbiosis: mutualism and communalism. Allanheld and Osmun, Montclair, NJ, pp 77–116

Weigelt A (1996) Experimentelle Untersuchungen zum Einfluß von Herbivorie durch Blattschneiderameisen auf die Sekundärstoffchemie beernteter Pflanzen und auf das Lichtklima im Einzelbaum an ausgewählten Arten des tropischen Regenwaldes von

Mittelamerika (Panama). Diploma Thesis, University of Würzburg, Würzburg, Germany

Welden CW, Hewett SW, Hubbel SP, Foster RB (1991) Sapling survival, growth and recruitment: relationship to canopy height in a neotropical forest. Ecology 72:35–50

Weller D (1986) Size related foraging in the leaf-cutting ant *Atta texana* (Buckley) (Formicidae: Attini). Funct Ecol 3:461–468

Welles JM, Norman JM (1991) Instrument for indirect measurement of canopy architecture. Agron J 83:818–825

Wetterer JK (1988) Size selective foraging by leaf-cutting ants and planktivorous fish. PhD Diss, University of Washington, Seattle

Wetterer JK (1990) Load-size determination in the leaf-cutting ant, *Atta cephalotes*. Behav Ecol 1:95–101

Wetterer JK (1991) Allometry and the geometry of leaf-cutting in *Atta cephalotes*. Behav Ecol Sociobiol 29:347–351

Wetterer JK (1994a) Attack by *Paraponera clavata* prevents herbivory by the leaf-cutting ant, *Atta cephalotes*. Biotropica 26:462–465

Wetterer JK (1994b) Forager polymorphism, size-matching, and load delivery in the leaf-cutting ant, *Atta cephalotes*. Ecol Entomol 19:57–64

Wetterer JK, Schultz TR, Meier R (1998) Phylogeny of fungus-growing ants (tribe Attini) based on mt DNA sequence and morphology. Mol Phylogenet Evol 9:42–47

Wheeler WM (1907) The fungus growing ants of North America. Bull Am Mus Nat Hist 23:669–807

Wheeler WM (1925) A new guest-ant and other Formicidae from Barro Colorado Island, Panama. Biol Bull Mar Biol Lab Woods Hole 49:150–181

White PS, Jentsch A (2001) The search for generality in studies of disturbance and ecosystem dynamics. In: Esser K, Lüttge U, Kadereit JW, Beyschlag W (eds) Progress in Botany, vol 62. Springer, Berlin Heidelberg New York, pp 399–449

White PS, Mackenzie MD, Busing RT (1985) A critique of overstory/understory comparisons based on transition probability analysis of an old growth spruce–fire stand in the Appalachians. Vegetatio 64:37–45

Whitehouse MEA, Jaffe K (1996) Ant wars: combat strategies, territory and nest defence in the leaf-cutting ant *Atta laevigata*. Anim Behav 51:1207–1217

Whitford KR, Colquhoun IJ, Lang ARG, Harper BM (1995) Measuring leaf area index in a sparse eucalypt forest: a comparison of estimates from direct measurement, hemispherical photography, sunlight transmittance and allometric regression. Agric For Meteorol 74:237–249

Whitham TG, Maschinski J, Larson KC, Paige KN (1991) Plant responses to herbivory: the continuum from negative to positive and underlying physiological mechanisms. In: Price PW, Lewinsohn TM, Wilson Fernandes G, Benson WW (eds) Plant-animal interactions: evolutionary ecology in tropical and temperate regions. Wiley, New York, pp 227–256

Wiens JA (2000) Ecological heterogeneity: an ontogeny of concepts and approaches. In: Hutchings MJ, John EA, Stewart AJA (eds) The ecological consequences of environmental heterogeneity. The 40th Symposium of the British Ecological Society held at the University of Sussex, 23–25 March 1999. Blackwell Science, Oxford, pp 9–31

Wills C (1996) Safety in diversity. New Sci 149:38–42

Wills C, Condit R, Foster RB, Hubbell SP (1997) Strong density- and diversity-related effects help to maintain tree species diversity in a neotropical forest. Proc Natl Acad Sci USA 94:1252–1257

Wilson EO (1971) The insect societies. Harvard University Press, Cambridge, MA

Wilson EO (1980a) Caste and division of labor in leaf-cutter ants (Hymenoptera: Formicidae: *Atta*). I. The overall pattern in *Atta sexdens*. Behav Ecol Sociobiol 7:143–156

Wilson EO (1980b) Caste and division of labor in leaf-cutter ants (Hymenoptera: Formicidae: *Atta*). II. The ergonomic organisation of leaf-cutting. Behav Ecol Sociobiol 7:157–165

Wilson EO (1983a) Caste and division of labor in leaf-cutter ants (Hymenoptera: Formicidae: *Atta*) III. Ergonomic resiliency in foraging by *A. cephalotes*. Behav Ecol Sociobiol 14:47–54

Wilson EO (1983b) Caste and division of labor in leaf-cutter ants (Hymenoptera: Formicidae: *Atta*). IV. Colony ontogeny of *A. cephalotes*. Behav Ecol Sociobiol 14:55–60

Wilson EO (1985) The sociogenesis of insect colonies. Science 228:1489–1495

Wilson EO (1986) The defining traits of fire ants and leaf-cutting ants. In: Lofgren CS, Vander Meer RK (eds) Fire ants and leaf-cutting ants. Westview Press, Boulder, CO, pp 1–9

Wilson EO (1990) Success and dominance in ecosystems: the case of social insects. Ecology Institute Oldendorf/Luhe, Oldendorf/Luhe

Wilson EO (1992) The diversity of life. WW Norton and Company, New York

Wilson SD (2000) Heterogeneity, diversity and scale in plant communities. In: Hutchings MJ, John EA, Stewart AJA (eds) The ecological consequences of environmental heterogeneity. The 40th Symposium of the British Ecological Society held at the University of Sussex, 23–25 March 1999. Blackwell Science, Oxford, pp 53–69

Windsor DM (1990) Climate and moisture variability in a tropical forest: long-term records from Barro Colorado Island, Panama. Smithson Contrib Earth Sci 29:1–145

Windsor DM, Rand AS, Rand WM (1990) Características de la precipitación en la isla de Barro Colorado. In: Leigh EG Jr, Rand AS, Windsor DM (eds) Ecología de un bosque tropical – ciclos estacionales y cambios a largo plazo. Smithsonian Institute Press, Washington, DC, pp 53–71

Wint GRW (1983) Leaf damage in tropical rain forest canopies. In: Sutton SL, Whitmore TC, Chadwick AC (eds) Tropical rain forest, ecology and management. Blackwell Scientific, Oxford, pp 229–239

Wirth R (1996) Der Einfluß von Blattschneiderameisen der Gattung *Atta* auf einen tropischen halbimmergrünen Regenwald in Panama: Herbivorieraten, Effekte auf das Lichtklima im Bestand und Ökologie der Futtersuche. PhD Diss, University of Würzburg, Würzburg, Germany

Wirth R, Beyschlag W, Ryel RJ, Hölldobler B (1997) Annual foraging of the leaf-cutting ant *Atta colombica* in a semideciduous rain forest in Panama. J Trop Ecol 13:741–757

Wirth R, Kost C, Schmidt S (2001a) Costs of trail construction and maintenance may affect the pattern of herbivory in leaf-cutting ants. In: Zotz G, Körner C (eds) The functional importance of biodiversity. Verh Ges f Ökologie, Parey, Berlin p 315

Wirth R, Weber B, Ryel RJ (2001b) Spatial and temporal variability of canopy structure in a tropical moist forest. Acta Oecol 22:235–244

Wolda H (1978) Fluctuations in abundance of tropical insects. Am Nat 112:1017–1045

Wong M, Wright SJ, Hubbell SP, Foster R (1990) The spatial pattern and reproductive consequences of outbreak defoliation in *Quararibea asterolepis*, a tropical tree. J Ecol 78:579–588

Woodring WP (1958) Geology of Barro Colorado Island, Canal Zone. Smithsonian Institution Miscellaneous Collections, vol 135. Smithsonian Institution, Washington, DC

Woodring WP (1964) Geology and paleontology of canal and adjoining parts of Panama. In: Reston VA (ed) US Geol Survey Prof Pap 306-A. US Geological Survey, Reston, VA

Wright SJ (2002) Plant diversity in tropical forests: a review of mechanisms of species coexistence. Oecologia 130:1–14

Yamakura T, Hagihara A, Sukardjo S, Ogawa H (1986) Aboveground biomass of tropical rain forest stands in Indonesian Borneo. Vegetatio 68:71–82

Yavitt JB (2000) Nutrient dynamics of soil derived from different parent material on Barro Colorado Island, Panama. Biotropica 32:198–207

Yavitt JB, Battles JJ, Lang GE, Knight DH (1995) The canopy gap regime in a secondary neotropical forest in Panama. J Trop Ecol 11:391–402

Yoda K (1974) Three-dimensional distribution of light intensity in a tropical rain forest of West Malaysia. Jpn J Ecol 24:247–254

Zamora R, Hodar JA, Gomez JM (1999) Plant-herbivore interaction: beyond a binary vision. In: Pugnaire FI, Valladares F (eds) Handbook of functional plant ecology. Marcel Dekker, New York, pp 677–718

Zeh JE, Zeh AD, Zeh DW (1999) Dump material as an effective small-scale deterrent to herbivory by *Atta cephalotes*. Biotropica 31:368–371

Zotz G (1992) Photosynthese und Wasserhaushalt von Pflanzen verschiedener Lebensformen des tropischen Regenwaldes auf der Insel von Barro Colorado, Panama. PhD Thesis, University of Würzburg, Würzburg, Germany

Zwölfer H (1987) Species richness, species packing, and evolution in insect-plant systems. In: Schulze ED, Zwölfer H (eds) Potentials and limitations of ecosystem analysis. Springer, Berlin Heidelberg New York, pp 301–319

Taxonomic Index

Page numbers in *italic* denote photographs and drawings.

Subject Index

Page numbers in *italic* denote photographs, drawings and maps.

Ecological Studies
Volumes published since 1997

Printing: Mercedes-Druck, Berlin
Binding: Stein + Lehmann, Berlin